$$F = \frac{9}{5}C + 32$$

$$C = \frac{5}{9}\left[F - 32\right]$$

Introduction to Polymer Viscoelasticity

Introduction to Polymer Viscoelasticity

Second Edition

John J. Aklonis
University of Southern California
Los Angeles, California

William J. MacKnight
University of Massachusetts
Amherst, Massachusetts

A Wiley-Interscience Publication
JOHN WILEY AND SONS
New York   Chichester   Brisbane   Toronto   Singapore

Library of Congress Cataloging in Publication Data:

Aklonis, John J.
    Introduction to polymer viscoelasticity.

    "A Wiley-Interscience publication."
    Includes bibliographies and index.
    1. Polymers and polymerization. 2. Viscoelasticity.
I. MacKnight, William J. II. Title.

TA455.P58A35   1983   620.1'9204232   82-17528
ISBN 0-471-86729-2

Printed in the United States of America

10 9 8 7 6 5 4 3 2 1

To A. V. Tobolsky who first introduced us to the mysteries of polymer viscoelasticity and to the meaning of scientific research.

Preface to the Second Edition

In the decade since the first edition of *Introduction to Polymer Viscoelasticity* appeared, we have noted a number of significant scientific developments. We also suffered a personal tragedy with the death of Professor M. C. Shen.

Among the major developments are a new approach to long-range relaxational motions known as the theory of reptation, and the further elucidation of the kinetic theory of rubber elasticity. In this second edition, we have attempted to take account of some of these developments on a level consistent with the introductory nature of the text. We have also added an entirely new chapter on dielectric relaxation, a technique now widely used to investigate molecular motions in polar polymers. Finally, we have tried to strengthen and clarify several other sections as well as eliminate errors or inconsistencies in the first edition that have been pointed out to us by colleagues and students.

Both of us felt very deeply the untimely death of Professor Shen, as did many others who valued his friendship and respected his scientific prowess. He made important and lasting contributions to such diverse areas as rubber elasticity theory, the understanding of mechanical properties of block copolymers, and plasma polymerization, to name but a few. His collaboration in the preparation of the second edition was sorely missed, but we feel that his influence remains clear and we are proud to acknowledge it.

We wish to thank Dr. Richard M. Neumann, who read the manuscript critically, and Ms. Teresa M. Wilder, who drew many of the figures. We are also indebted to Dr. Neumann and Professor L. L. Chapoy for furnishing some of the new problems contained in this edition.

Once again we accept sole responsibility for any errors in the text, be they old ones remaining from the first edition or new ones that may appear in the second.

J. J. AKLONIS
W. J. MACKNIGHT

Los Angeles, California
Amherst, Massachusetts
December 1982

Preface to the First Edition

The viscoelastic response of polymeric materials is a subject which has undergone extensive development over the past twenty years and still accounts for a major portion of the research effort expended. It is not difficult to understand the reason for this emphasis in view of the vast quantities of polymeric substances which find applications as engineering plastics and the still greater volume which are utilized as elastomers. The central importance of the time and temperature dependence of the mechanical properties of polymers lies in the large magnitudes of these dependencies when compared to other structural materials such as metals. Thus an understanding of viscoelastic behavior is fundamental for the proper utilization of polymers.

Viscoelasticity is a subject of great complexity fraught with conceptual difficulties. It is possible to distinguish two basic approaches to the subject which we shall designate as the continuum mechanical approach and the molecular approach. The former attempts to describe the viscoelastic behavior of a body by means of a mathematical schema which is not concerned with the molecular structure of the body, while the latter attempts to deduce bulk viscoelastic properties from molecular architecture. The continuum mechanical approach has proven to be very successful in treating a large number of problems and is of very great importance. However, it is not our intention to treat this approach rigorously in this text. Rather, we shall be concerned with the molecular approach and attempt to present a basic foundation upon which the reader can build. The fundamental difficulty encountered with the molecular approach lies in the fact that polymeric materials are large molecules of very complex structures. These structures are too complex, even if they were known in sufficient detail which, in general, they are not, to lend themselves to mathematical analysis. It is therefore necessary to resort to simplified structural models, and the results deduced from these are limited by the validity of the models adopted.

Several excellent treatments of molecular viscoelasticity are available. (See the references of Chapter 1.) The book by Professor Ferry, in particular, is an exhaustive and complete exposition. The question may then be asked, why the necessity for still another text and one restricted to bulk amorphous polymers, at that? Such a question must send each of the authors scurrying in quest of an "apologia pro vita sua." The answer to the question lies in the use of the word "introduction" in the title. What we have attempted to do is to provide a detailed grounding in the fundamental concepts. This means, for example, that all derivations have been presented in great detail, that concepts and models have been presented with particular attention to assumptions, simplifications, and limitations, and that problems have been provided at the end of each chapter to illustrate points in the text. The level of mathematical difficulty is such that the average baccalaureate chemist should be able to readily grasp it. Where more advanced mathematical techniques are required, such as transform techniques, the necessary methods are developed in the text.

Having attempted to delineate what this book is, it may be well to remind the reader what it is not. First of all, it is not a complete treatment —lacking among other topics discussions of crystalline polymers, solution behavior, melt rheology, and ultimate properties. It is also not written from the continuum mechanics approach and thus is not mathematically sophisticated. Finally, it is not a primer of polymer science. Familiarity with the basic concepts of the field is presumed.

The authors' first acquaintance with the literature of viscoelastic behavior in polymers evoked a response much like that experienced by neophytes in the literary arts on a first reading of "Finnegan's Wake," by James Joyce. It is immediately apparent that one is in the presence of a great work, but somehow it will be necessary to master the language before appreciation, let alone understanding, may be achieved. Recognizing the nature of the problem, Joycean scholars came to the rescue with works of analysis to provide a skeleton key to "Finnegan's Wake." Proper utilization of this skeleton key will open the door to an understanding of that forbidding masterwork. It was thus our intent to provide a similar skeleton key to the literature of molecular viscoelasticity. How well we have succeeded must be left to the judgment of our readers.

We are grateful to the students at our respective institutions who have suffered our attempts to present the material in this text in coherent form at various stages of development. Their criticisms and suggestions have led to significant improvements. We are also grateful to Mrs. William Jackson, who translated many rough sketches into finished drawings. It is hardly to be expected that a work of this nature could be free from errors. We have attempted to eliminate as many as possible but, of course, bear full responsibility for those remaining.

JOHN J. AKLONIS
WILLIAM J. MACKNIGHT
MITCHEL SHEN

Los Angeles, California

Contents

Introduction to
Polymer
Viscoelasticity

1

Introduction

The subject matter of this book is the response that certain classes of substances make when they are subjected to external forces of various kinds. The classes of substances to be treated are known as viscoelastic bodies. As the name implies, these materials respond to external forces in a manner intermediate between the behavior of an elastic solid and a viscous liquid. In order to set the stage for what follows, it is necessary to quantitatively define the types of force to which the viscoelastic bodies are subjected.

Consider first the motion of a rigid body in space. This motion can be thought of as consisting of translational and rotational components. If no forces act on the body, it will maintain its original state of motion indefinitely in accordance with Newton's first law of motion. However, if a single force or a set of forces whose vector sum is nonzero act on the body, it will experience acceleration or a change in its state of motion. Consider, however, the case where the vector sum of forces acting on the body is zero and the body experiences no change in either its translational or rotational component of motion. In such a condition, the body is said to be *stressed*. If the requirement of rigidity is removed, the body will in general undergo a deformation as a result of the application of these balanced forces. If this occurs, the body is said to be *strained*. It is the relationship between stress and strain which is our main concern. Depending on the types of stress and strain applied to a body, it is possible to use these quantities to define new quantities that relate to the mechanical strength of the body. These new quantities are called moduli. In order to understand the physical meaning of the modulus of a solid, consider the following simple experiment.

Suppose we have a piece of rubber (e.g., polyisobutylene) $\frac{1}{2}$ cm $\times \frac{1}{2}$ cm $\times$ 4 cm and a piece of plastic (e.g., polystyrene) of the same dimensions. The

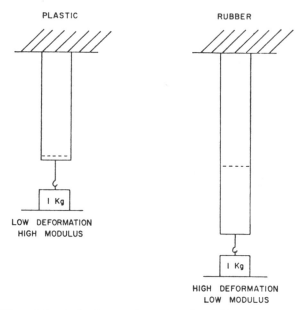

Figure 1-1. A simple extension test used to define the modulus.

experiment to be performed consists of suspending a weight (applying a force) of, say 1 kg, from each material as shown in Figure 1-1. As is obvious, the deformation of the rubber will be much greater than that of the plastic. Using this experiment we define a modulus as a quantity proportional to the applied force divided by the change in length:

$$M \propto \frac{\text{Force}}{\Delta L} \tag{1-1}$$

Since ΔL is much larger for the rubber than for the plastic, from equation (1-1) it is clear that the modulus of the rubber is much lower than the modulus of the plastic. Thus the particular modulus defined in equation (1-1) defines the resistance of a material to elongation.

Further experimentation, however, reveals that the situation is more complicated than is initially apparent. If, for example, one were to carry out the test on the rubber at liquid nitrogen temperature, one would find that this "rubber" undergoes a much smaller elongation than with the same force at room temperature. In fact, the extension would be so small as to be comparable to the extension exhibited by the plastic at room temperature. A more dramatic demonstration of this effect is obtained by immersing a rubber ball in liquid nitrogen for several minutes. The cold ball, when bounced, no longer has the characteristic properties of a rubbery object but, instead, is indistinguishable from a hard sphere made of plastic.

On the other hand, if the piece of plastic is heated in an oven to 130°C and then subjected to the modulus measurement, it is found that a much larger elongation, comparable to the elongation of the rubber at room temperature, results.

These simple experiments indicate that the modulus of a polymeric material is not invariant, but is a function of temperature, that is, $M = M(T)$.

An investigation of the temperature dependence of the modulus of our two samples is now possible. At temperature T_1 we measure the modulus as before, then increase the temperature to T_2, and so on. Schematic data from such an experiment are plotted in Figure 1-2. The temperature dependence of the modulus is so great that it must be plotted on a logarithmic scale. (This great variation in modulus presents experimental problems which will be treated subsequently.) The modulus at the lowest temperature measurement (plastic) has been arbitrarily defined as 1.0 in the figure, and other moduli are given relative to this value. The area between

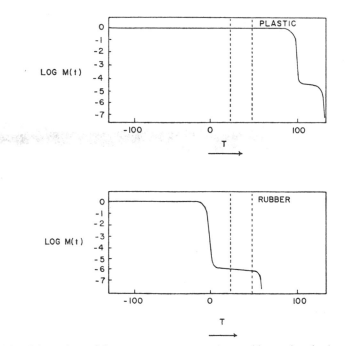

Figure 1-2. Schematic modulus–temperature curves for a rubber and a plastic over a broad temperature range.

the vertical dashed lines represents room temperature and, consistent with the first experiment, we find that in this range the plastic has a high modulus while the rubber has a relatively low modulus. Upon cooling, the modulus of the rubber rises markedly, by about four orders of magnitude, indicating that the rubber at lower temperatures behaves like a plastic. Another drastic change in modulus for the rubber is evidenced at higher temperatures; here the material is becoming softer, as indicated by further decreases in modulus. This behavior is discussed in detail in Chapter 3. The modulus–temperature behavior for the plastic is seen to be quite similar to that of the rubber except that the changes occur at higher temperatures, resulting in the high modulus observed at room temperature. At 135°C it is clear that the modulus of this material is that of a rubber, agreeing with the results of one of the earlier "experiments" in this discussion.

One more type of deformational experiment remains to be discussed. Consider a material like pitch or tar, which is used as a roof coating and is applied at elevated temperatures. Our test is similar to the standard experiment done above, utilizing the same size sample at room temperature. First we suspend the 1-kg weight from the sample and observe the small resultant extension. According to equation (1-1), the modulus calculated is high. However, if the weight is removed and the sample left suspended in this vertical position for several days, it is observed that the rather weak force exerted by gravity results in a considerable elongation of the sample. Now application of equation (1-1) gives a very low value for the modulus. Thus the modulus measurement on the short time-scale of minutes resulted in a high value while the modulus measurement on the longer time-scale of days resulted in a low modulus. This apparent discrepancy is accounted for by realizing that the modulus is a function of time as well as temperature; this will be found to be the case generally when considering polymeric systems. Strictly then, the measurements spoken of earlier in this chapter and depicted schematically in Figure 1-2 should have some time associated with each modulus value. (Time represents the duration between the application of the force and the measurement of the extension fraction.) It is convenient to pick the same constant time for all measurements so one might consider the constant time-factor in Figure 1-2 to be 10 seconds.

As is evident from the above discussion, it should be possible to measure the behavior of a material as a function of time at constant temperature. A schematic modulus–time behavior is shown in Figure 1-3. Here again the maximum value of the modulus is normalized to 1.0 (log 1.0 = 0). The modulus is seen to fall from its initial high value by about three orders of magnitude to a modulus indicative of a rubber and, after evidencing a plateau, fall again. The ordinate here is $\log t$; at this temperature an experiment lasting for 1 to 30 minutes would characterize this material as a

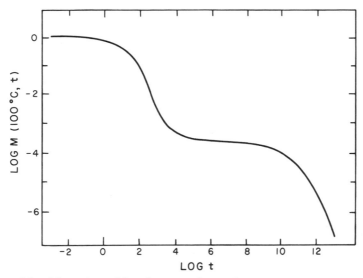

Figure 1-3. Schematic modulus–time curve for a polymer at constant temperature.

plastic. However, to an experiment lasting 10^8 minutes (200 years), the material would "look like" a rubber. Longer measurements would correspond to still softer materials. Methods for obtaining curves of the type shown in Figure 1-3 are discussed in Chapter 3, as well as methods of converting from modulus–time behavior to modulus–temperature behavior and vice versa.

Another experiment is often carried out in laboratories dealing with the physical properties of polymers. This is the determination of the temperature at which the material properties change from those of a plastic to those of a rubber. This temperature is known as the glass transition temperature and is a characteristic property of each substance. In Figure 1-2, for example, it is clear that at about 100°C, the modulus of the plastic exhibits a steep decrease. Careful analysis of the curve in this region, however, indicates no abrupt change in modulus but rather a smoothly varying change. From this experiment, it would seem that the glass transition occurs over a range of temperatures rather than at a single temperature. Experimentally, it has been found that the coefficient of expansion of a substance undergoes a more abrupt change in the region of the glass transition. The temperature at this change, in fact, is defined as the glass transition temperature T_g.

An example of the data obtained in the determination of a glass transition temperature is presented in Figure 1-4. The volume of a sample is measured as a function of temperature, care being taken to change the

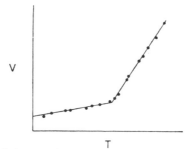

Figure 1-4. Schematic of data used to determine the glass transition temperature of a polymer.

temperature slowly and at an essentially constant rate. Experimentally, one often employs a technique which is based on Archimedes' Principle. The temperature where there is a change in slope (due to a discontinuity in the coefficient of expansion) is taken as T_g.

SUGGESTED READING

Alfrey, T., *Mechanical Behavior of High Polymers*, Interscience, New York, 1948.

da C. Andrade, E. N., *Viscosity and Plasticity*, Chemical Publishing Co., New York, 1951.

Bueche, F., *Physical Properties of Polymers*, Interscience, New York, 1962.

Christensen, R. M., *Theory of Viscoelasticity, An Introduction*, Academic Press, New York, 1971.

Eirich, F. R., Ed., *Rheology: Theory and Applications*, Vol. 1, Academic Press, New York, 1956.

Ferry, J. D., *Viscoelastic Properties of Polymers*, 3rd ed., Wiley, New York, 1980.

Flüger, W., *Viscoelasticity*, 2nd ed., Springer-Verlag, New York, 1975.

Gross, B., *Mathematical Structure of the Theories of Viscoelasticity*, Hermann, Paris, 1953.

McCrum, N. G., B. E. Read, and G. Williams, *Anelastic and Dielectric Effects in Polymeric Solids*, Wiley, New York, 1967.

Middleman, S., *The Flow of High Polymers*, Interscience, New York, 1968.

Nielsen, L. E., *Mechanical Properties of Polymers*, Rheinhold, New York, 1962.

Pipkin, A. C., *Lectures on Viscoelasticity Theory*, Applied Mathematical Sciences, Vol. 7, Springer-Verlag, New York, 1972.

Timoshenko, S., and J. N. Goodier, *Theory of Elasticity*, McGraw-Hill, New York, 1951.

Tobolsky, A. V., *Properties and Structure of Polymers*, Wiley, New York, 1960.

Tobolsky, A. V., and H. F. Mark, Eds., *Polymer Science and Materials*, Wiley, New York, 1971.

Varga, O. H., *Stress–Strain Behavior of Elastic Materials*, Interscience, New York, 1966.

Ward, I. M., *Mechanical Properties of Solid Polymers*, Wiley, New York, 1971.

Zener, C., *Elasticity and Anelasticity of Metals*, University of Chicago Press, Chicago, 1948.

2

Phenomenological Treatment of Viscoelasticity

In order to put the concepts discussed in the Introduction on a more quantitative basis, and to better understand some of the added complications that arise from the time dependence of the modulus, several methods commonly used to make physical measurements on polymeric systems are discussed in this chapter. First, however, it is necessary to rigorously define the parameters to be derived from the experimental data.

A. ELASTIC MODULI

Perhaps the simplest deformation that can be applied to a sample is uniaxial tension or compression, shown in Figure 2-1a. This is the type of deformation mentioned in the Introduction. However, our previous concept of force is modified slightly in this more quantitative discussion to become the stress[†] defined as

$$\sigma_t = \frac{F}{AB} \tag{2-1}$$

[†]In this context, both stress and strain are defined considering very small deformations. Under these conditions, for example, the cross-sectional area of the sample before the deformation is virtually equal to that after the deformation. For large deformations, "true" and "engineering" functions are defined in texts such as references 1 and 2.

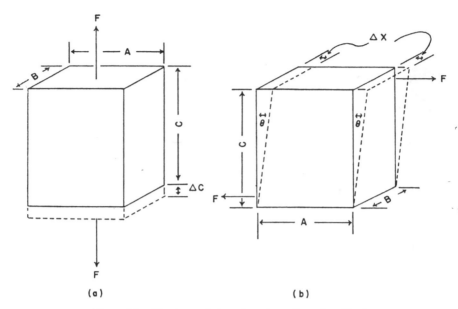

Figure 2-1.   Tension and shear in a three-dimensional body.

where the subscript t represents uniaxial tension and the other symbols are defined in the figure. σ_t has the units of dynes per cm^2, newtons per $meter^2$, or PSI. Note that static definitions are being considered here, that is, the parameters do not vary with time. These initial considerations will shortly be generalized to time-dependent functions.

The application of a stress to a real body will result in a strain, as discussed in the Introduction. The fractional extension is defined as the tensile strain. Thus ε, the tensile strain resulting from the application of a uniaxial stress, is given as

$$\varepsilon = \frac{\Delta C}{C} \tag{2-2}$$

and it is clear that ε is dimensionless. In terms of classical physics, the tensile modulus E and the tensile compliance D are defined by equation (2-3).

$$E = \frac{\sigma_t}{\varepsilon} = \frac{1}{D} \tag{2-3}$$

Again it must be stressed that this discussion is strictly limited to time-independent phenomena.

The second type of deformation is illustrated in Figure 2-1b and is called simple shear. The application of the force F will result in the deformation shown by the dashed lines. Here the stress σ_s is

$$\sigma_s = \frac{F}{AB} \qquad (2\text{-}4)$$

where the subscript s denotes shear. The shear strain is given as

$$\gamma = \frac{\Delta X}{C} = \tan \theta \qquad (2\text{-}5)$$

where θ is the angle shown and is related to $\Delta X / C$ by simple trigonometry. The shear modulus G and the shear compliance J are defined using the relationship

$$G = \frac{\sigma_s}{\gamma} = \frac{1}{J} \qquad (2\text{-}6)$$

From the theory of elasticity of isotropic solids, one has the relationship between E and G and D and J that

$$E = 2(1+\mu)G \quad \text{or} \quad J = 2(1+\mu)D \qquad (2\text{-}7)$$

where μ is Poisson's ratio defined as

$$\mu \equiv \frac{1}{2}\left[1 - \left(\frac{1}{V}\right)\frac{dV}{d\varepsilon}\right] \qquad (2\text{-}8)$$

Here, the fundamental relationship between shear and tensile properties is demonstrated. Equation (2-7) is derived in Appendix 1 of this chapter. For the special case of an incompressible material, $(dV/d\varepsilon)=0$, $\mu=0.5$, and one obtains the results

$$E = 3G \quad \text{or} \quad 3D = J$$

Experimentally, μ is quite close to 0.5 for a rubber but is lower, in the range of 0.2 to 0.3, for some plastics and still lower for certain heterogeneous materials.

B. TRANSIENT EXPERIMENTS

Consideration of the time dependence of relaxation phenomena adds additional complications. The value of a measured modulus or compliance will very definitely depend on the exact manner in which the experiment is carried out. As an example, consider the following experiments. First, a polymer is subjected to a constant uniaxial stress σ_1 for one hour; this perturbation results in some measurable strain, say ε (1 hour). In a second experiment, however, an identical sample is subjected to sufficient stress to result in the same strain ε (1 hour) immediately upon application of the stress. Then the stress is decreased so that the strain remains constant at ε (1 hour). The value of the stress after 1 hour in the second experiment is defined as σ_2. In general σ_1 and σ_2 will not be the same, the stress associated with the constant strain experiment being lower. However, since the strains are the same, the two "modulus" values calculated from equation (2-3) are different. Consequently one must explicitly state the method in which a parameter is to be measured in its definition. Fortunately, physics allows us to relate most parameters obtained from different experiments so that a knowledge of some of them will suffice to define all the others.[3-7] Some of these transformations will be treated later in this chapter.

In Chapter 1, we dealt with hanging a weight on a sample and observing the extension. This is a crude form of a creep experiment. In such an experiment the sample is subjected to constant stress and its strain is measured as a function of time. The shear creep compliance $J(t)$ is defined as

$$J(t) = \frac{\gamma(t)}{\sigma_{s_0}} \qquad (2\text{-}9)$$

where σ_{s_0} is the constant shear stress and the tensile creep compliance

$$D(t) = \frac{\varepsilon(t)}{\sigma_{t_0}} \qquad (2\text{-}10)$$

where σ_{t_0} is the constant tensile stress.

Experimentally one might fasten a cubic sample to a light aluminum plate and to a support as shown in Figure 2-2. The sample would then be placed into a constant temperature environment where it would be allowed to come to thermal equilibrium before the start of the experiment (provision having been made to eliminate the stress due to the unloaded weight pan and string). When the experimenter is satisfied that the sample is at equilibrium, he places the appropriate weights on the weight pan and observes the deformation of the sample by noting the position of the weight pan with respect to the scale behind the apparatus. It should be clear that the shear strain $\gamma(t)$ as a function of time is easily calculated from these

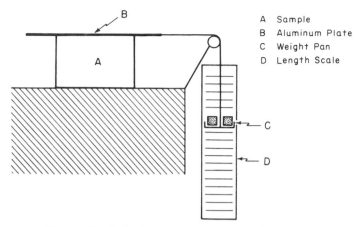

A   Sample
B   Aluminum Plate
C   Weight Pan
D   Length Scale

Figure 2-2. A simple apparatus to measure shear creep.

data and a knowledge of the original sample dimensions. The force exerted by the weights is used to calculate the constant shear stress and then the shear creep compliance $J(t)$ is calculated using the expression (2-9). It must be emphasized that this experimental setup is among the crudest imaginable and will give correspondingly imprecise results unless considerable refinement is made. The apparatus of Plazek[8] represents what is presently among the most sophisticated instrumentation used for measuring creep compliances.

Another common type of experiment, called stress relaxation, may be considered. Here as in the experiment above, the strain is maintained constant and the stress is measured as a function of time. A simple stress relaxation balance is shown in Figure 2-3. The sample, in the form of a strip, is first attached to the upper clamp and then the forces on either side of the fulcrum are balanced by adjusting the counter weight (B). Next the lower clamp is attached and the sample raised to the temperature desired and thermal equilibrium established. (Usually the sample is raised to a temperature about 10 degrees above the test temperature or to a temperature above the glass temperature to allow possible "locked in" strains to relax before approach to the test temperature is made). At the start of the experiment, the strain is introduced quickly via the micrometer with the balance beam jammed against the upper stop. The application of weight to the balance pan will eventually cause the beam to move down from the upper stop toward the lower stop. This is the equilibrium point when the forces on either side of the fulcrum are balanced and simple calculations yield the stress at nearly constant strain. (The strain varies *minutely* during

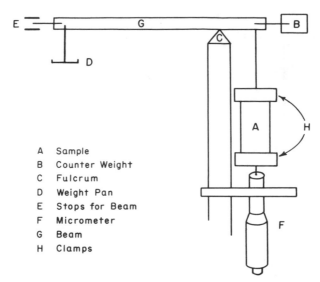

A   Sample
B   Counter Weight
C   Fulcrum
D   Weight Pan
E   Stops for Beam
F   Micrometer
G   Beam
H   Clamps

Figure 2-3.   A simple apparatus to measure tensile stress relaxation.

the measurement.) This experiment measures $E(t)$, the tensile relaxation modulus,

$$E(t) = \frac{\sigma_t(t)}{\varepsilon_0} \tag{2-11}$$

with ε_0 being the constant tensile strain.

Similarly, a shear stress relaxation experiment would measure $G(t)$, the shear stress relaxation modulus.

$$G(t) = \frac{\sigma_s(t)}{\gamma_0} \tag{2-12}$$

with γ_0 being the constant shear strain.

This experiment has also been depicted in a rather simplified form. Usually the mechanical mechanism is supplanted by electronics which, in essence, perform the same tasks.

Equations of the form of (2-3) and (2-6) relating moduli and compliances are no longer applicable to our new time-dependent functions, since

$$E(t) = \frac{\sigma_t(t)}{\varepsilon_0} \neq \frac{\sigma_{t_0}}{\varepsilon(t)} = \frac{1}{D(t)} \tag{2-13}$$

as was discussed in the example mentioned earlier in this section.

Other moduli and compliances can be defined and they are not, in general, equal to the functions defined above. Remember that $G(t)$ and $E(t)$ can only be measured directly from constant strain experiments, while $J(t)$ and $D(t)$ can only be measured directly from constant stress experiments. This fact is sometimes forgotten and can easily lead to considerable error.

C. DYNAMIC EXPERIMENTS

In addition to creep and stress relaxation experiments, another type of measurement is quite common. Here the stress or strain, instead of being a step function, is an oscillatory function with an angular frequency ω. Dynamic modulus values measured using such perturbations are functions of ω rather than time. The problem of putting dynamic experiments on a quantitative level is only slightly more difficult than is the case with step deformation experiments. In order to become familiar with this situation, we will begin by considering a simple experiment.

First, suppose that a rod-shaped sample of a completely elastic substance is fixed in a chuck and rotated at some frequency ω. Then by means of a suitable bearing device, we hang a weight W_0 from the end of the rod as shown in Figure 2-4. We know from experience that the rod will deform in the manner shown in Figure 2-5a (i.e., "straight down"). Formally, it may be said that stress and strain remain in phase regardless of the frequency and independently of the time elapsed since the start of the experiment.

A short analysis of the stress function will be helpful. In normal laboratory coordinates, we consider the force to be "downward" (stationary) and for the sample to be rotating at frequency ω. Considering the reference system to be the sample instead of the laboratory reveals,

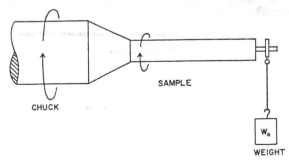

Figure 2-4. A novel dynamic experiment. After B. Maxwell, *J. Polym. Sci.*, **20** 551 (1956).

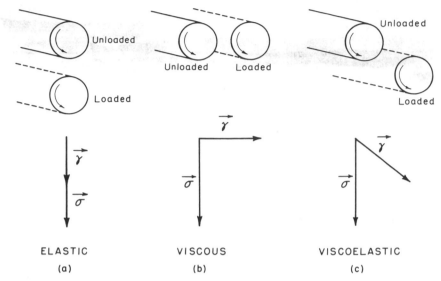

ELASTIC   VISCOUS   VISCOELASTIC

(a)   (b)   (c)

Figure 2-5. Actual displacements of a rod subjected to the dynamic experiment shown in Figure 2-4: (a) an elastic rod, (b) a viscous rod, (c) a viscoelastic rod.

however, that this is the same as having the sample stationary and the weight rotating about the sample at frequency ω. This is just another view of the same experiment. At some time in the rotation of the weight all of the stress σ_0 is in the *down* direction. As the weight moves up through the cycle, the stress in the *down* direction decreases until, after 90° of rotation, the stress in this direction is zero. The stress continues to decrease and becomes $-\sigma_0$ at 180° from down, that is, up. The stress application in any direction can be treated as the projection of a vector of length σ_0 rotating at frequency ω on an axis in the direction of the stress being considered. The projection of a rotating vector on an axis yields a sine-shaped function; the stress application in any direction of the sample is thus sinusoidal as a function of time. The situation is shown in Figure 2-6a.

For an elastic rod the modulus is *not* time dependent. Thus, assuming the deformation is in shear, we may write:

$$\gamma(t) = \frac{\sigma(t)}{G} \tag{2-14}$$

where the time dependence comes exclusively from the nature of the perturbation. The stress is just given by the function

$$\sigma(t) = \sigma_0 \cos \omega t \tag{2-15}$$

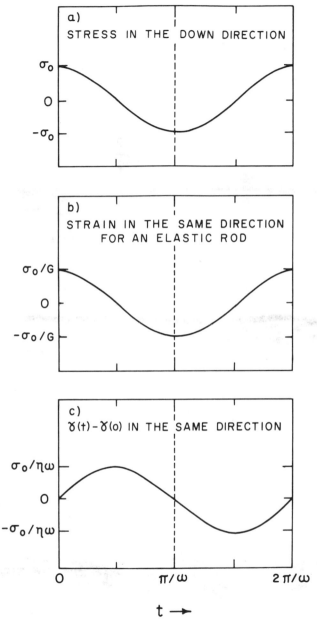

Figure 2-6. Stress and strain in a dynamic experiment.

which then yields the strain as

$$\gamma(t) = \frac{\sigma_0}{G}\cos \omega t \tag{2-16}$$

This behavior is shown in Figure 2-6b.

Secondly, consider a perfectly viscous rod. The fundamental behavior is described by the equation

$$\frac{d\gamma}{dt} = \frac{\sigma(t)}{\eta} \tag{2-17}$$

The strain rate, $d\gamma/dt$, responds linearly to the stress for a viscous body with a proportionality constant η, called the viscosity coefficient or just the viscosity.

In equation (2-17) it is clear that the rate of strain is a maximum when the stress is a maximum and a minimum when the stress is a minimum. Since a sine wave has its maximum rate of change when it passes through zero amplitude and its minimum rate of change when the amplitude is a maximum, and since the stress in our experiment is sinusoidal, the stress and the strain must be 90° out of phase for a sample whose properties are defined by equation (2-17).

Mathematically, if $\sigma(t)$ is given by equation (2-15), equation (2-17) then becomes

$$\frac{d\gamma}{dt} = \frac{\sigma_0}{\eta}\cos \omega t \tag{2-18}$$

which can be integrated to yield

$$\gamma(t) - \gamma(0) = +\frac{\sigma_0}{\eta\omega}\sin \omega t \tag{2-19}$$

This equation is plotted in Figure 2-6c and the configuration of a perfectly viscous rod subjected to these experimental conditions is that shown in Figure 2-5b.

Finally, the general case of a viscoelastic rod will result in a displacement somewhat between the two extremes; that is, the strain will lag the stress by between 0° and 90° (Figure 2-5c).

It is not often that a dynamic experiment is carried out with the stress vector immobile in the laboratory coordinates and the sample rotating. Usually the sample is stationary in the laboratory and responds to an oscillatory stress or strain. Both of these situations are identical, however,

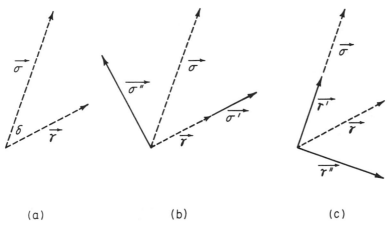

Figure 2-7.   (a) Orientation of stress and strain vectors in a dynamic experiment. (b) Stress decomposed. (c) Strain decomposed.

except for the experimenter's point of view and from both of them he can measure the same physical properties.

Now consider a sample of viscoelastic material subjected to a sinusoidal shear stress. This is easily visualized using the technique of rotating vectors. In Figure 2-7 the magnitude of the vector σ represents the maximum of the stress applied to our sample. It rotates counterclockwise at an angular frequency ω. The actual stress on the sample is represented by the projection of this vector onto a suitable axis such as the y axis. Thus as σ rotates, its projection on the y axis is the sinusoidal stress experienced by the sample. As was mentioned previously, the strain will lag the stress by some amount, which is usually called the loss angle δ. In the rotating vector scheme, the strain is represented, as shown in Figure 2-7, by the vector γ, which rotates at the same frequency as σ with a magnitude proportional to the maximum strain. The loss angle δ is the angle separating γ and σ. The strain in the sample, likewise, is represented by the projection of the strain vector on the same axis as was used for the stress vector projection. It is obvious from the figure that the maximum in stress does not coincide with the maximum in strain.

The absolute shear modulus $|G|$ is defined as the magnitude of the stress vector divided by the magnitude of the strain vector; the absolute shear compliance $|J|$ is the reciprocal of this quantity.

Quite often it is convenient to separate the viscoelastic response into "in-phase" and "out-of-phase" components. This is done in Figure 2-7. Here the projection of σ onto γ yields σ', the component of σ in phase with

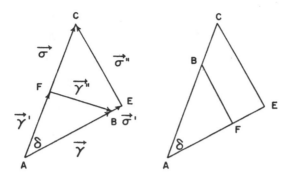

σ	AC	σ''	CE
γ	AB	γ'	AF
σ'	AE	γ''	BF

Figure 2-8.   Rearrangement of Figure 2-7.

the strain, while the projection of σ on the axis perpendicular to γ yields σ'', the component 90° out of phase with the strain. The in-phase and out-of-phase shear moduli, G' and G'', are given as[†]

$$G' = \frac{\sigma'}{\gamma} \qquad G'' = \frac{\sigma''}{\gamma} \tag{2-20}$$

Similarly, the projection of γ onto σ yields γ' the strain component in phase with the stress; the projection of γ onto a vector axis perpendicular to σ yields γ'', the out-of-phase component of the strain. Here the shear compliance functions J' and J' are defined as

$$J' = \frac{\gamma'}{\sigma} \qquad J'' = \frac{\gamma''}{\sigma} \tag{2-20a}$$

In Figure 2-8 the vectors are rearranged to make it clear that

$$\tan\delta = \frac{\sigma''}{\sigma'} = \frac{J''}{J'} = \frac{\gamma''}{\gamma'} = \frac{G''}{G'} \tag{2-21}$$

[†]Boldface type indicates vectors. Italic type indicates magnitudes.

where $\tan \delta$ is often called the loss tangent. Parameters with one prime are called storage functions and those with two primes loss functions. This has to do with the fact that in-phase stress and strain results in elastically stored energy which is completely recoverable, whereas one-fourth-cycle-separated stress and strain results in the dissipation of energy, which is lost to the system (Problem 2-4).

In the above considerations, a sinusoidal shear stress was applied to the sample. It should be clear that a sinusoidal tensile stress is equally applicable and that definitions of $E'(\omega)$, $E''(\omega)$, $|D|$, D', and so on would yield results completely analogous to the derived shear parameters.

The nomenclature of complex moduli and compliances is also often used. Here the out-of-phase component is made the imaginary part of a complex parameter; thus the complex shear modulus $G*$ and the complex shear compliance $J*$ are defined as

$$G* = G' + iG''$$
$$J* = J' - iJ'' \tag{2.22}$$

The use of complex numbers to represent the functions has no particular significance except that by using the complex plane, one has a simple method for representing orthogonal vectors.

D. BOLTZMANN SUPERPOSITION PRINCIPLE

The Boltzmann superposition principle is one of the simplest but most powerful principles of polymer physics.[10] We have previously defined the shear creep compliance as relating the stress and strain in a creep experiment.†

$$\gamma(t) = \sigma_0 J(t) \tag{2-23}$$

The stress σ_0 is applied instantaneously at time equal to zero. One might, however, imagine an experiment where the stress σ_1 is applied, not at $t = 0$, but at some other arbitrary time, perhaps u_1. Then equation (2-23) would become

$$\gamma(t) = \sigma_1 J(t - u_1) \tag{2-24}$$

---

†The subscripts t and s will be dropped from σ. J and G will be considered to result from σ_s. D and E will be considered to result from σ_t. All arguments presented concerning shear variables are completely applicable to tensile parameters also.

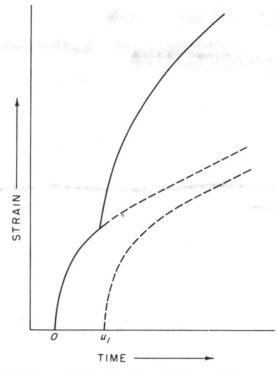

Figure 2-9. Linear addition of strains applied sequentially to a sample.

Consider now the application of two stress increments σ_0 and σ_1 at the times $t = 0$ and $t = u_1$ respectively. The Boltzmann superposition principle asserts that the two stresses act independently and the resultant strains add linearly. This situation is shown in Figure 2-9. Thus

$$\gamma(t) = \sigma_0 J(t) + \sigma_1 J(t - u_1) \tag{2-25}$$

or for a more general experiment consisting of discrete stress increments $\sigma_1, \sigma_2, \ldots, \sigma_n$ applied at times $t = u_1, u_2, \ldots, u_n$

$$\gamma(t) = \sum_{i=1}^{n} \sigma_i J(t - u_i) \tag{2-26}$$

The summation of the individual σ_i's would represent the total stress so that in considering a continuous stress application, $\sigma(t)$, the increment of applied stress is just the derivative of $\sigma(t)$. Replacing the summation by an integration and remembering that u is the variable results in

$$\gamma(t) = \int_{-\infty}^{t} \frac{\partial \sigma(u)}{\partial u} J(t-u)\,du \qquad (2\text{-}27)$$

Note that in this equation t has become the *fixed* time of the observation of the strain. The stress history is accounted for in terms of the integration variable u. The limits of integration are taken as $-\infty$, since the complete stress history contributes to the observed strain, and t since it is obvious that stresses applied after t, the time of observation of the strain, can have no effect on the observed strain.

In a completely analogous manner, one may derive an expression relating the stress $\sigma(t)$ to the strain in a sample that has experienced some continuous strain history given by the function $\gamma(t)$:

$$\sigma(t) = \int_{-\infty}^{t} \frac{\partial \gamma(u)}{\partial u} G(t-u)\,du \qquad (2\text{-}28)$$

Equations (2-27) and (2-28) are often given in an alternative form, which we will now derive. Integrating equation (2-27) by parts

$$\int w\,dv = -\int v\,dw + \int d(wv) \qquad (2\text{-}29)$$

where

$$dv = \left(\frac{\partial \sigma(u)}{\partial u}\right) du \qquad w = J(t-u)$$

one obtains

$$\gamma(t) = J(t-u)\sigma(u)\big|_{-\infty}^{t} - \int_{-\infty}^{t} \sigma(u)\frac{\partial J(t-u)}{\partial u}\,du \qquad (2\text{-}30)$$

We assume that $\sigma(-\infty)$ is equal to zero. Setting $t-u$ equal to a, a new variable, and observing new limits of integration due to this variable change, gives

$$\gamma(t) = J(0)\sigma(t) + \int_{0}^{\infty} \sigma(t-a)\frac{\partial J(a)}{\partial a}\,da \qquad (2\text{-}31)$$

the strain at time (t) and
the recoverable compliance

In an analogous manner, equation (2-28) becomes

$$\sigma(t) = G(0)\gamma(t) + \int_0^\infty \gamma(t-a)\frac{\partial G(a)}{\partial a}\,da \qquad (2\text{-}32)$$

As a specific example of the use of the Boltzmann principle, consider a material with a creep compliance given by the function

$$J(t) = J_r + t/\eta \qquad (a)$$

where J_r represents the recoverable deformation and η is the viscosity. Let us calculate the strain $\gamma(t)$ when this body is subjected to the linear stress function $\sigma(t)$ shown in part (a) of Figure 2-10; (i) at the time t_1 during the loading and (ii) at the time t_2 after the loading has ceased.

i.   For $t = t_1$, the stress is given, in terms of its variable (u), as

$$-\infty \leqslant u \leqslant 0 \qquad \sigma(u) = 0$$
$$0 \leqslant u \leqslant t_1 \qquad \sigma(u) = ku$$

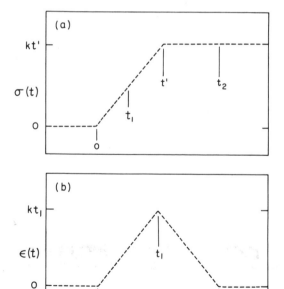

Figure 2-10.   Stress and strain functions used in calculations.

Making use of equation (2-27), one has

$$\gamma(t_1) = \int_0^{t_1} k\left(J_r + \frac{t_1}{\eta} - \frac{u}{\eta}\right) du \tag{b}$$

and carrying out this simple integration leads to

$$\gamma(t_1) = kt_1 J_r + \frac{kt_1^2}{2\eta} \tag{c}$$

Remembering that kt_1 is just the total applied stress at the time t_1 yields

$$\gamma(t_1) = \sigma(t_1)\left(J_r + \frac{t_1}{2\eta}\right) \tag{d}$$

Equation (d) then gives an expression for the strain at time t_1 in a body whose creep compliance is given by equation (a) when it is subjected to a linear loading pattern starting at time zero. Note that t_1 must occur during the stressing period.

ii. We now calculate the strain after the stress addition has stopped. Again we can summarize the stressing history:

$$-\infty \leqslant u \leqslant 0 \qquad \sigma(u) = 0$$

$$0 \leqslant u \leqslant t' \qquad \sigma(u) = ku$$

$$t' \leqslant u \leqslant t_2 \qquad \sigma(u) = kt'$$

Now equation (2-27) becomes

$$\gamma(t_2) = \int_0^{t'} k\left[J_r + \frac{t_2}{\eta} - \frac{u}{\eta}\right] du \tag{e}$$

Again, this is a simple integration which gives the result:

$$\gamma(t_2) = \sigma(t')\left[J_r + \frac{t_2}{\eta} - \frac{t'}{2\eta}\right] \tag{f}$$

It is interesting to carry out part (**ii**) of this example using the more widely used equation (2-31) instead of equation (2-27). One must use the new variable in the integral so that the strain history is introduced in terms of this transformed variable rather than in terms of the normal laboratory time. Writing equation (2-31) for part (**ii**) of the example gives

$$\gamma(t_2) = J(0)\sigma(t_2) + \int_0^\infty \sigma(t_2 - a)\frac{\partial J(a)}{\partial a}\,da \tag{g}$$

In laboratory time, there was no stress imposed between the time $= -\infty$ and $= 0$. In terms of the variable a, however, this corresponds to $a = \infty$ and $a = t_2$ for when $a = \infty$, $t_2 - a = -\infty$ and when $a = t_2$, $t_2 - a = 0$. This is a consequence of the variable change used to derive equation (2-31). Application of this equation without changing variables will necessarily lead to an incorrect result. Completing the stress summary in the usual way:

$$-\infty \leqslant u \leqslant 0 \qquad \sigma(t) = 0 \qquad \infty \geqslant a \geqslant t_2$$

$$0 \leqslant u \leqslant t' \qquad \sigma(t) = ku \qquad t_2 \geqslant a \geqslant t_2 - t'$$

$$t' \leqslant u \leqslant t_2 \qquad \sigma(t) = kt' \qquad t_2 - t' \geqslant a \geqslant 0$$

From equation (a)

$$\frac{\partial J(a)}{\partial a} = \frac{1}{\eta} \tag{h}$$

Substituting equation (h) and the strain history into equation (g) yields

$$\gamma(t_2) = J_r\sigma(t') + \int_0^{t_2 - t'} \frac{k(t')}{\eta}\,da + \int_{t_2 - t'}^{t_2} \frac{k}{\eta}(t_2 - a)\,da \tag{i}$$

Integration and cancellation gives (f) as indeed it must.

Any of equations (2-27), (2-28), (2-31), or (2-32) is completely sufficient as a statement of the Boltzmann superposition principle. Often in particular applications, however, it is more convenient to use one form than another.

E.   RELATIONSHIP BETWEEN THE CREEP COMPLIANCE AND THE STRESS RELAXATION MODULUS

One of the direct consequences of the Boltzmann superposition principle is that there is a relationship between the stress relaxation modulus and the creep compliance. We have already seen that when dealing with time-independent functions, the compliance and the modulus are simply the reciprocals of each other. This simple relationship no longer holds in the time-dependent case. Fortunately, the Boltzmann superposition principle allows us to express moduli and compliances in terms of one another even in the time-dependent case. We will derive these relationships directly from equations (2-31) and (2-32) with the aid of Laplace transforms. A short introduction or review of Laplace transform techniques will be presented first.

Transform techniques in general are extremely powerful mathematical tools. The manner in which a transform operates is to take a problem in equation form from one space, where its solution is difficult, to another space where, it is hoped, the solution will be simpler. The solution in transform space is then transformed back into the original space to yield the answer to the problem. The Laplace transform of a function $F(t)$ is denoted $f(p)$ or $L(F(t))$ and is defined as

$$f(p) \equiv L(F(t)) \equiv \int_0^\infty e^{-pt} F(t)\, dt \tag{2-33}$$

It is informative to derive several relationships among Laplace transforms that will be used later. These relationships are to be found in any table of Laplace transforms.

First consider the calculation of the Laplace transform of the function

$$F(t) = at \tag{2-34}$$

where a is a constant.

Substitution into equation (2-33) gives

$$L(at) = \int_0^\infty e^{-pt} at\, dt \tag{2-35}$$

which upon integration yields

$$L(at) = \frac{a}{p^2} \tag{2-36}$$

Thus it is clear that

$$L(t) = \frac{1}{p^2} \qquad L(aF(t)) = aL(F(t)) \tag{2-37}$$

Next, consider the Laplace transform of the function $F(t-a)$, where $F(x) = 0$ for $x < 0$. Again, substitution into equation (2-33) gives

$$L(F(t-a)) = \int_0^\infty e^{-pt} F(t-a)\, dt \tag{2-38}$$

Now letting $t - a = x$, one has

$$L(F(t-a)) = e^{-ap} \int_0^\infty e^{-xp} F(x)\, dx = e^{-ap} L(F(t)) \tag{2-39}$$

Lastly, consider the transform of $F'(t)$. Proceeding as above,

$$L(F'(t)) = \int_0^\infty e^{-pt} F'(t)\, dt \tag{2-40}$$

Integration by parts yields

$$L(F'(t)) = e^{-pt} F(t) \big|_0^\infty + p \int_0^\infty e^{-pt} F(t)\, dt \tag{2-41}$$

The second term is just the definition of the Laplace transform of $F(t)$ times p; evaluation of the first term at the limits of integration gives

$$L(F'(t)) = -F(0) + pL(F(t)) \tag{2-42}$$

It is clear that this expression is very powerful, allowing one to find the transform of the derivative of a function from a knowledge of the function itself; it is not necessary to know the derivative of the function explicitly.

One additional result is needed, and it is derived in Appendix 2 of this chapter. This is Borel's theorem, which states:

$$L\left(\int_0^t F_1(t-\tau) F_2(\tau)\, d\tau \right) = L(F_1(t)) L(F_2(t)) \tag{2-43}$$

Using these results, we can now derive the relationship between the creep compliance and the stress relaxation modulus.

The Laplace transform of equation (2-31) yields

$$L(\gamma(t)) = J(0)L(\sigma(t)) + \int_0^\infty e^{-pt} \int_0^\infty \sigma(t-a) \frac{\partial J(a)}{\partial a}\, da\, dt \tag{2-44}$$

or

$$L(\gamma(t)) = J(0)L(\sigma(t)) + \int_0^\infty \frac{\partial J(a)}{\partial a} \int_0^\infty e^{-pt} \sigma(t-a)\, dt\, da \tag{2-45}$$

Making use of the result derived in equation (2-39) gives

$$L(\gamma(t)) = J(0)L(\sigma(t)) + \left[\int_0^\infty e^{-ap} \frac{\partial J(a)}{\partial a}\, da \right] L(\sigma(t)) \tag{2-46}$$

The term enclosed in brackets, however, is nothing except the Laplace transform of the derivative of $J(t)$. Thus, we may apply the result obtained in equation (2-42) to get

$$L(\gamma(t)) = J(0)L(\sigma(t)) + L(\sigma(t))\left[pL(J(t)) - J(0) \right] = pL(\sigma(t))L(J(t))$$

$$(2\text{-}47)$$

Next, transform equation (2-32) into Laplace space in the same manner to obtain

$$L(\sigma(t)) = G(0)L(\gamma(t)) + L(\gamma(t))\left[pL(G(t)) - G(0) \right]$$

$$= pL(\gamma(t))L(G(t)) \qquad (2\text{-}48)$$

Equations (2-47) and (2-48) give

$$\frac{1}{p^2} = L[G(t)]L[J(t)] \qquad (2\text{-}49)$$

Relationship between creep compliance and the stress relaxation modulus

This is the solution of the problem in transform space. We have a direct relationship between the transforms of the compliance and the modulus. This solution must now be returned to real space. Making use of Borel's theorem, equation (2-43), and the result derived in equation (2-37) gives the final result:

$$t = \int_0^t G(\tau)J(t-\tau)\,d\tau \qquad (2\text{-}50)$$

This is the convolution integral that is the relationship between the creep compliance and the stress relaxation modulus. It is exact and depends only on the applicability of the Boltzmann superposition principle.

These equations are used to convert a modulus to a compliance in problem 6 of this chapter. The results of this calculation are depicted in Figure 2-11 (see reference 11).

F. RELATIONSHIP BETWEEN STATIC AND DYNAMIC PROPERTIES

Still another relationship between experimental parameters is a direct consequence of the Boltzmann superposition principle. We will derive the equations relating the tensile stress relaxation modulus $E(t)$ to the in-phase and out-of-phase dynamic tensile moduli $E'(\omega)$ and $E''(\omega)$, starting from

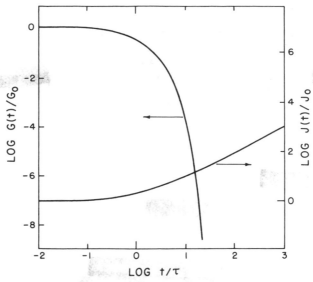

Figure 2-11. Relationship between a creep compliance and a stress relaxation modulus (special case).

equation (2-28) expressed in tensile quantities. Equation (2-28) may be written as

$$\sigma(t) = \int_{-\infty}^{t} \frac{\partial \varepsilon(u)}{\partial u} E(t-u)\, du \qquad (2\text{-}51)$$

A simple variable change setting $s = t - u$ yields

$$\sigma(t) = -\int_{0}^{\infty} \frac{\partial \varepsilon(t-s)}{\partial s} E(s)\, ds \qquad (2\text{-}52)$$

Consider the application of a sinusoidal strain, which may be represented by

$$\varepsilon(t) = \varepsilon_0 e^{i\omega t} = \varepsilon_0(\cos \omega t + i \sin \omega t) \qquad (2\text{-}53)$$

where ε_0 is the maximum amplitude of the strain. Substitution of equation (2-53) into equation (2-52) and simplification gives

$$\frac{\sigma(t)}{\varepsilon(t)} = \int_{0}^{\infty} i\omega e^{-i\omega s} E(s)\, ds \qquad (2\text{-}54)$$

By expanding the term $e^{-i\omega s}$, as shown in equation (2-53), one obtains

$$E^*(\omega) = \int_0^\infty \omega(\sin \omega s) E(s)\, ds + i \int_0^\infty \omega(\cos \omega s) E(s)\, ds \qquad (2\text{-}55)$$

Here we have observed that the ratio of the tensile stress to the tensile strain in an experiment where the perturbing function may be given a complex sinusoidal representation is the complex tensile modulus $E^*(\omega)$. From this expression, one immediately realizes that either of the two dynamic modulus functions may be calculated from the tensile relaxation modulus as follows:

$$E'(\omega) = \omega \int_0^\infty \sin \omega s E(s)\, ds$$

$$\qquad (2\text{-}56)$$

$$E''(\omega) = \omega \int_0^\infty \cos \omega s E(s)\, ds$$

This is a specific example illustrative of the method by which one may relate static and dynamic properties. Listings of other results may be found in texts such as Ferry.[3]

It is apparent that equation (2-56) embodies a Fourier sine and cosine transformation of $E(t)$; thus normal Fourier transform[12] methods permit the inversion of these relations to give the static modulus as a function of the dynamic properties.

Also in this section we have tacitly assumed that $E(t)$ approaches zero at long times. If this is not true, as is the case with a crosslinked polymer, one may introduce the function $E(t) - E_e$ where E_e represents the equilibrium tensile modulus. It is clear that this difference does indeed become zero at long times.

APPENDIX 1:   RELATIONSHIP BETWEEN TENSILE AND SHEAR PARAMETERS

$$\mu \equiv \frac{1}{2}\left[1 - \frac{1}{V}\left(\frac{dV}{d\varepsilon}\right)\right] \qquad (a)$$

Consider the application of a small tensile stress to a sample as shown in Figure 2-1a. The volume will change by some amount ΔV under this perturbation and the tensile strain will change from 0 to ε. Thus equation

(a) becomes

$$\mu = \frac{1}{2}\left(1 - \frac{1}{\varepsilon}\frac{\Delta V}{V}\right) \tag{b}$$

Since ε is small, however, the fractional volume change is just $\varepsilon - 2d$ where d is $|\Delta A/A|$ or $|\Delta B/B|$ (A and B both decrease). This result obtains, since for the figure shown,

$$\frac{\Delta V}{V} \approx \frac{AB\Delta C + AC\Delta B + BC\Delta A}{ABC}$$

$$\approx \frac{\Delta C}{C} + \frac{\Delta B}{B} + \frac{\Delta A}{A} = \varepsilon - 2d \tag{c}$$

Substituting equation (c) into equation (b) yields the simple result

$$\mu = \frac{d}{\varepsilon} \tag{d}$$

which is an alternate definition of the Poisson ratio. From here on we will use the proof outlined in reference 1. First we must show that a shear strain γ can be replaced by two tensile strains each of magnitude $\gamma/2$.

In Figure 2-12 we show a cross-section of a cube $ABCD$ that is sheared as shown. In addition, several diagonals have been drawn. Here the shear strain γ is clearly $\tan\theta$ and equal to CC'/CB.

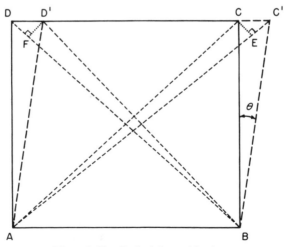

Figure 2-12. Body deformed in shear.

Construct the line CE, which is perpendicular to the rhombus diagonal AC'. Since the strain is small, $\angle CC'E$ is close to $\pi/4$ and $C'E = CC'/\sqrt{2}$. Similarly, $AE \approx AC$. Now define the tensile strain along the diagonal AC' as

$$\varepsilon = \frac{C'E}{AE} = \frac{CC'/\sqrt{2}}{AC} = \frac{CC'/\sqrt{2}}{CB\sqrt{2}}$$

or   (e)

$$\varepsilon = \frac{1}{2}\frac{CC'}{CB} = \frac{\gamma}{2}$$

Similarly, we could show that a negative strain of magnitude $\gamma/2$ results from consideration of the motion of corner D. Thus a shear strain can be decomposed into two perpendicular tensile strains, each half the magnitude of the tensile strain, of opposite sign and both oriented 45° to the shear strain.

Thus let linear stresses σ_t and $-\sigma_t$ which are perpendicular to each other act on a sample to generate a shear strain that is oriented 45° away from both of these forces. In the direction of $+\sigma_t$, the tensile strain ε is just

$$\varepsilon = \frac{\sigma_t}{E} + \mu\frac{\sigma_t}{E}$$ (f)

The first term on the right side arises from the fact that the sample elongates in the direction of the $+\sigma_t$ application because of the $+\sigma_t$ application; simultaneously, it elongates even more due to the application of the $-\sigma_t$ stress perpendicular to the $+\sigma_t$ direction. From equations (d) and (f):

$$\mu\frac{\sigma_t}{E} = \frac{d}{\varepsilon}\frac{\sigma_t}{E} = \frac{d}{\varepsilon}\varepsilon = d$$ (g)

Since the tensile strain is just $\frac{1}{2}$ the magnitude of the shear strain, equation (f) becomes

$$\gamma = \frac{2\sigma_t}{E}(1+\mu)$$ (h)

Finally, we must show that a tensile stress of magnitude σ_t acting in the x direction and a tensile stress of magnitude σ_t acting in the Z direction is equivalent to a shear stress σ_s of magnitude σ_t acting in a direction 45° to

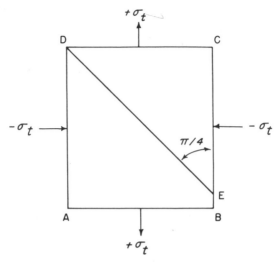

Figure 2-13.   Shear as a result of tensile stresses.

each tensile stress. To do this, consider Figure 2-13, where these stresses are shown. This is just a cross-section of a three-dimensional cube. The actual magnitude of the force on the surface (surface area S) represented by the line BC is just $S\sigma_t$ and is numerically equal to the magnitude of the forces on the plane represented by the line DC. If the stress is infinitesimal, the points B and E become coincident and the resultant of the forces mentioned above is just ($\sqrt{2}\ S\sigma_t$) and directed *along* the new plane represented by the line DB. The shear stress along this plane, then, is just obtained by dividing the force along the plane by the area it is acting upon to give (area of the plane represented by DB is $\sqrt{2}\ S$):

$$\sigma_s = \frac{\sqrt{2}\ S\sigma_t}{\sqrt{2}\ S} = \sigma_t \tag{i}$$

By definition, however,

$$\sigma_s = G\gamma \tag{j}$$

Nevertheless, according to equations (h) and (i),

$$\sigma_t = G\frac{2\sigma_t}{E}(1+\mu)$$

which can be rearranged to give

$$E = 2G(1+\mu) \tag{k}$$

APPENDIX 2:   BOREL'S THEOREM

Consider the two functions $F_1(t)$ and $F_2(t)$ whose Laplace transforms are given as $f_1(p)$ and $f_2(p)$ respectively. Then the product

$$f_1(p)f_2(p) = \int_0^\infty F_1(T)e^{-pT} \int_0^\infty e^{-pt}F_2(t)\, dt\, dT \qquad \text{(a)}$$

If

$$F_2(t) = 0 \qquad \text{for} \quad t < 0 \qquad \text{(b)}$$

however, application of equation (2-39) yields

$$f_1(p)f_2(p) = \int_0^\infty F_1(T) \int_0^\infty e^{-pt}F_2(t-T)\, dt\, dT \qquad \text{(c)}$$

Upon rearrangement we get

$$f_1(p)f_2(p) = \int_0^\infty e^{-pt} \int_0^\infty F_2(T)F_2(t-T)\, dT\, dt \qquad \text{(d)}$$

and remembering condition (b) yields

$$f_1(p)f_2(p) = L\left[\int_0^t F_1(T)F_2(t-T)\, dT \right]$$

PROBLEMS

1. Calculate the weight needed to bring about a shear deformation, ΔX, of 0.40 cm, of a viscoelastic cube 2 cm on a side after 10^{-4}, 10^{-2}, 10^0, 10^4, and 10^6 seconds. The time-dependent shear compliance is

$$J(t) = \left(10^{-9} + \frac{t}{10^9} \right) \ Pa^{-1} \qquad (Pa = newton/m^2)$$

2. Make a table expressing J', J'', $\tan \delta$, $|J|^2$, $|G|^2$, G', and G'' in terms of (a) G' and G'', (b) J' and J'', (c) $|G|$ and $\tan \delta$, and (d) $|J|$ and $\tan \delta$.

3. Calculate the stress at time t longer than $2t_1$ in a body with a relaxation modulus given by

$$E(t) = E_1 e^{-t/\tau_1} + E_2 e^{-t/\tau_2}$$

when the body is subjected to the strain history depicted in Figure 2-10b.

4. By considering a sinusoidal stress applied to a viscoelastic body, show that in-phase strain results in conservation of energy (work) whereas out-of-phase strain results in energy dissipation.

5. Show that

$$J(t) = \frac{\sin m\pi}{m\pi} \frac{1}{G(t)}$$

if $\log J(t) = \log A + m \log t$. Note $\Gamma(n)\Gamma(1-n) = \pi/\sin(n\pi)$ where $\Gamma(n)$ is the gamma function defined as

$$\int_0^\infty x^{n-1} e^{-x} \, dx \equiv \Gamma(n) \quad \text{and} \quad \Gamma(n) = (n-1)! \quad \text{for} \quad n > 0$$

6. Calculate the creep compliance for a body whose stress relaxation modulus is given by

$$G(t) = G_0 e^{-t/\tau}$$

7. Calculate $G'(\omega)$ and $G''(\omega)$ for a body whose stress relaxation modulus is given by the expression in Problem 2-6.

8. The strain in a dynamic shear experiment may be represented as

$$\gamma^* = \gamma_0 e^{i\omega t}$$

Since, in general the stress will lead the strain, the corresponding function for the stress is

$$\sigma^* = \sigma_0 e^{i(\omega t + \delta)}$$

Obtain expressions for $G'(\omega)$ and $G''(\omega)$ using this formalism.

9. The behavior of viscoelastic materials subjected to oscillatory perturbations may also be treated by generalizing the concept of viscosity (rather than modulus) and separating it into in-phase and out-of-phase components. Thus Newton's law (see Chapter 7, Section A) becomes

$$\sigma^* = \eta^* \frac{d\varepsilon^*}{dt}$$

where $\eta^* = \eta' - i\eta''$. Here η' measures energy dissipation, and η'' measures stored energy. In this formalism, show that

$$\eta'(\omega) = \frac{G''(\omega)}{\omega} \quad \text{and} \quad \eta''(\omega) = \frac{G'(\omega)}{\omega}$$

10. Show that the form of the Boltzmann principle given in equation (2-27) reverts to the defining equation for the shear creep compliance, equation (2-9), when a sample, initially at rest, is subjected to an instantaneous increment of stress at $t = 0$, which is thereafter held constant.

It may be helpful to note that the Dirac delta function $\delta(a)$ can be defined as

$$f(x) = \int_{-\infty}^{\infty} \delta(x - x')f(x')\,dx'$$

That is, it has unit area when $a = 0$ and is zero everywhere else.

11. Stress–strain curves are often measured by monitoring the tensile stress as a sample, originally at rest, is subjected to a constant tensile strain *rate* starting at $t = 0$. Show that, at any subsequent time during the constant strain-rate period, the slope of the stress–strain curve is the tensile stress relaxation modulus:

$$E(t) = \frac{\dot{\sigma}}{\dot{\varepsilon}}$$

12. Suppose that a material with a tensile stress relaxation modulus given as

$$E(t) = E_0 e^{-t/\tau}$$

is used in an ordinary stress relaxation experiment starting at $t = 0$ and employing a constant strain ε_0. At some later time, t', the stress is suddenly removed. Show that the strain at times greater than t' is

$$\varepsilon(t) = \varepsilon_0(1 - e^{-t'/\tau})$$

REFERENCES

1. R. B. Lindsay, *Physical Mechanics*, Van Nostrand, New York, 1933, pp. 317–324.

2. L. Malvern, *Introduction to the Mechanics of a Continuous Medium*, Prentice-Hall, New York, 1969.

3. J. D. Ferry, *Viscoelastic Properties of Polymers*, 3rd ed., Wiley, New York, 1980.

4. A. V. Tobolsky, *Properties and Structure of Polymers*, Wiley, New York, 1960.

5. F. Bueche, *Physical Properties of Polymers*, Interscience, New York, 1962.

6. B. Gross, *Mathematical Structure of the Theories of Viscoelasticity*, Hermann, Paris, 1953.

7. F. R. Eirich, Ed., *Rheology: Theory and Application*, Vol. 1, Academic Press, New York, 1956.

8. D. J. Plazek, *J. Polym. Sci., Part A-2*, **6**, 621 (1968).

9. B. Maxwell, *J. Polym. Sci.*, **20**, 551 (1956).

10. H. Leaderman, *Elastic and Creep Properties of Filamentous Materials*, The Textile Foundation, Washington, D.C., 1943.

11. I. L. Hopkins and R. W. Hamming, *J. Appl. Phys.*, **28**, 906 (1957).

12. S. Goldman, *Transformation Calculus and Electrical Transients*, Prentice-Hall, New York, 1949.

3

Time — Temperature Correspondence

Having examined some of the basic manifestations of the phenomenological aspects of viscoelasticity, we now shift our emphasis to interpretations of this behavior on a molecular scale. The concepts of the previous chapter are strictly independent of the existence of molecules; the results of those concepts fall into the realm of continuum mechanics. In this chapter we observe the results of generalized experiments and then interpret specific behavioral patterns in terms of the accepted molecular mechanisms. Furthermore, the basis of ideas concerning time–temperature correspondence is introduced.

A. FOUR REGIONS OF VISCOELASTIC BEHAVIOR

Once again, begin by considering an experiment. A polymer sample of unit cross-sectional area is subjected to an instantaneous tensile strain which is thereafter maintained constant. The stress is monitored as a function of time, and the tensile stress relaxation modulus is obtained using equation (2-11). Let t in equation (2-11) be any arbitrary time, perhaps 10 seconds. Next the stress is removed, allowing the sample to relax, and the temperature is changed. The same experiment is carried out yielding E (10 seconds) at this new temperature. The experiment is repeated at many temperatures to yield the "ten-second tensile relaxation modulus" as a function of temperature.

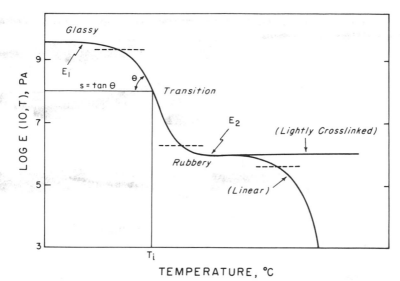

Figure 3-1. Schematic modulus–temperature curve showing various regions of viscoelastic behavior.

The types of behavior most often observed are shown in Figures 3-1 to 3-3. (We have selected tensile relaxation as an example; other types of deformation would have been equally suitable.)

Figure 3-1 shows idealized modulus–temperature curves for typical linear and crosslinked amorphous polymers. In this plot, four regions of viscoelastic behavior are defined. At low temperatures where the modulus is higher than 10^9 Pa, the polymer is hard and brittle; this is the glassy region. The glassy modulus, E_1, a slowly decreasing function of temperature, is a useful parameter to use in characterizing polymeric behavior. In this glassy region, thermal energy is insufficient to surmount the potential barriers for rotational and translational motions of segments of the polymer molecules. The chain segments are essentially "frozen" in fixed positions on the sites of a disordered quasi-lattice with their segments vibrating around these fixed positions much like low-molecular-weight molecules in a molecular crystal. With increasing temperature, the amplitude of vibrational motion becomes greater, and eventually the thermal energy becomes roughly comparable to the potential energy barriers to segment rotation and translation. In this temperature region, the polymer is at the glass transition temperature where short-range diffusional motions begin. Segments are free to "jump" from one lattice site to another; the brittle glass becomes a resilient leather.

This phenomenon is accompanied by a catastrophic decrease in the modulus of several decades, as indicated in Figure 3-1. The breadth of this transition region ranges from 5 to more than 20 degrees centigrade, depending on the nature of the polymer in question. The transition region may be characterized by two parameters. The temperature at which the 10-second modulus reaches 10^8 Pa is termed the inflection temperature, T_i, and the negative slope of the curve ($\tan \theta$) at this point is called s. Values of these characteristic viscoelastic parameters for several amorphous polymers are listed in Table 3-1.

As the temperature is further increased, the modulus again reaches a plateau region. This rubbery plateau is characterized by the modulus E_2 as shown in Figure 3-1. In this temperature interval, the short-range diffusional motions of the polymer segments that initially gave rise to the glass transition occur very much faster than our measurement time of 10 seconds. On the other hand, the long-range cooperative motion of chains that would result in translational motions of complete molecules is still greatly restricted by the presence of strong local interactions between neighboring chains. In the case of the crosslinked material, these interactions consist of primary chemical bonds. In the linear polymer, they are known as entanglements, and their precise nature is not clear. In any case, in the rubbery plateau region, segments of chains reorient relative to each other but large-scale translational motion does not occur.

The viscoelastic responses of linear and crosslinked polymers through the rubbery plateau region are essentially identical. As the temperature is further increased, however, differences between these two categories of polymers become evident, as shown in Figure 3-1. First consider a crosslinked network. As temperature is increased, the crosslinks consisting of primary chemical bonds remain intact, preventing the chains from translating relative to one another. Thus, although the modulus changes slightly with temperature in the rubbery plateau region of a crosslinked polymer (Chapter 6, Section A), this is a small effect compared to changes like those exhibited during the glass transition. Thus, to a first approximation, the modulus will remain constant for a crosslinked rubber up to temperatures where chemical degradation begins to occur.

The situation is quite different for a linear polymer. In this case, increasing temperature causes molecular motions to become more and more large-scale until eventually whole polymer molecules begin to translate. When the temperature is high enough, local chain interactions are no longer of sufficiently high energy to prevent molecular flow. During the 10-second test, the molecules will slip by one another, releasing much of the local strain, and the polymer sample will exhibit a correspondingly low

Table 3-1. Characteristic Viscoelastic Parameters for Selected Polymers[1]

Polymer	$3\,G_1$ (Pa)	$3G_2$ (Pa)	T_i (°C)	s (°C^{-1})
Crystalline Polymers				
Polyethylene				
($M_w = 2.6 \times 10^5$)	3×10^9	3×10^6	75	0.01
Polyvinyl chloride	4	2	78	0.25
Amorphous Linear Polymers				
Polyisobutylene	3.5	1	−62	0.15
Natural rubber				
(unvulcanized)	2.5	4	−67	0.2
Polystyrene	2	0.5	101	0.2
Polymethyl acrylate	3	1.5	16	0.2
Polymethyl methacrylate	1.5	2	107	0.15
Polybutyl acrylate	1.5	0.5	−53	0.2
Polybutyl methacyrlate	1	1	31	0.15
cis-Polybutadiene	2	1	−106	0.2
Polyethylene				
tetrasulfide	2	4	−24	0.15
Atactic polypropylene	2	2.5	−16	0.2
Polyacenaphthalene	—	—	264	0.1
Polyethylhexyl acrylate	—	0.5	−70	0.2
Selenium	6	—	45	0.15
Sulfur	—	—	−28	0.2
Bisphenol A polycarbonate				
($M_w = 4 \times 10^5$)	1.5	5	150	0.3
Polyethyl ether disulfide	0.5	2	−53	0.25
Polyethyl formal				
disulfide	1	0.5	−58	0.35
Amorphous Lightly Crosslinked Polymers				
Natural rubber (vulcanized)	3.5	4	−57	0.2
Polytetrahydrofuran	3	10	−73	0.15
Poly-2-hydroxylethyl				
methacrylate	3	2.5	96	0.1
Poly-*t*-butylaminoethyl				
methacrylate	2.5	1	41	0.1
Polyethyl methacrylate	2.5	2.5	77	0.1
Poly-*n*-propyl				
methacrylate	2	2	56	0.1
Polymethoxyethyl				
methacrylate	2	1.5	23	0.1
Polydiethylaminoethyl				
methacrylate	2	0.5	26	0.1

stress value. Since the molecules are translating relative to one another in this region, it should be expected that a considerable amount of flow will result from experiments carried out under these conditions. When the strain is released in our stress relaxation experiment, the sample will not recover to its former length but will relax to a new equilibrium state having a greater length. Thus, during the experiment the sample has undergone flow, and this area of the 10-second modulus temperature curve is therefore called the flow region.

If temperature is increased still further, barring chemical reaction the sample will become a viscous liquid. Our simple experimental setup is no longer applicable at high temperatures where the sample will not support its own weight. The modulus will, however, continue to decrease.

With this background, it will be easy to proceed with the analysis of the modulus–temperature behavior of more complicated systems.

First consider the effect of different molecular weights in linear amorphous systems. Figure 3-2 shows the relaxation behavior of three polystyrene samples. Sample (a) has a very low molecular weight, in the neighborhood of 10,000, while samples (b) and (c) have molecular weights of 1.4×10^5 and 2.17×10^5 respectively. First consider samples (b) and (c). At temperatures below 120°C, the materials are identical. There is no differentiation between the two samples since short-range segmental motions are determining the behavior. Such motions are essentially independent of

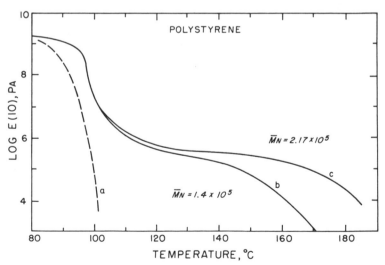

Figure 3-2. Ten-second modulus versus temperature curve for polystyrene of various molecular weights.

molecular weight at high molecular weights. At higher temperatures, however, where translational motions of the complete molecules determine the
relaxation behavior of the sample, it is most reasonable to expect longer
molecules to maintain a pseudonetwork to higher temperatures. This is
indeed observed as shown in Figure 3-2. Sample (a), however, appears to
behave anomalously. We notice that the initial decay of the glassy modulus
occurs at lower temperatures than is the case with the high molecular
weight polymers and, moreover, there is no rubbery plateau. The behavior
of sample (a) in the glass transition region may be qualitatively accounted
for on the basis of "free volume" considerations, as follows (more detail is
given in Chapter 4). Since a linear polymer chain has two ends, decreasing
the length of the chains leads to an increase in the concentration of ends in
the sample. However, since a chain middle is attached to other segments on
both sides, it is less mobile than a chain end attached on only one side.
Thus a polymer sample that has a greater concentration of chain ends than
another comparable sample will exhibit greater chain mobility at a given
temperature. In other words, a decrease in molecular weight leads to a
decrease in glass transition temperature. The absence of a rubbery plateau
is due to the fact that the short molecules present in this low molecular
weight sample are not long enough to form an "entanglement" network,
which is necessary for the observation of a rubbery state. (There appears to
be a critical molecular weight necessary for an entanglement network to
form.)

The above considerations must be modified rather extensively for the
case of a crystalline polymer. Figure 3-3 shows the 10-second modulus–

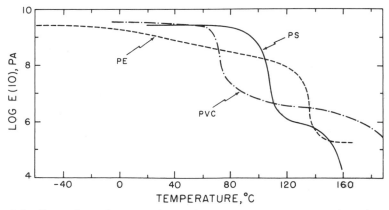

Figure 3-3. Comparison of ten-second modulus versus temperature curves for polystyrene,
polyvinyl chloride and polyethylene.

temperature curve for two typical crystalline polymers, polyethylene (PE) and polyvinyl chloride (PVC), contrasted with atactic polystyrene (PS), a typical noncrystalline polymer.

It should be emphasized that a crystalline polymer is not totally crystalline but partly crystalline and partly amorphous. Thus one expects to see behavior characteristic of the crystalline regions superposed upon behavior characteristic of the amorphous regions. This superposition is not necessarily a linear one but may rather be a complex coupling of the response of each network. Furthermore, the degree of crystallinity as well as the crystallite size is not a unique feature of any polymer system. Both of these properties are determined by prior thermal history.

The general characteristics attributed to the amorphous network are not greatly modified in crystalline polymers. There is still a glass-to-rubber transition, a rubbery plateau, and finally, in noncrosslinked systems, a flow region. This amorphous network, however, couples with the crystalline portion of the polymer. The main transition exhibited by the crystalline part of the polymer is a melting from an ordered crystal to a liquidlike, disordered state. The temperature of this melting is usually called T_m. With these thoughts in mind, one may now analyze the curves in Figure 3-3. Consider polyethylene. T_m is about 125°C. While the temperature of the glass transition is not yet known with certainty, it seems clear that it is low. At temperatures between the glass transition temperature and T_m, the polymer exhibits a very high modulus and experimentally polyethylene is also noted to have a very high impact strength in this region. Molecularly, this is to be expected since, in polyethylene, a large portion of the polymer is crystalline. Therefore, in this region one is observing the relaxation of a small amount of material held together by numerous hard crystallites. Here a high modulus should be expected. As T_m is approached, changes in the crystalline superstructure and in the degree of crystallinity result in a rapid decrease of the modulus. At the melting point, the modulus drops sharply to the rubbery modulus of polyethylene. With the crystallites gone, we are left with essentially a normal amorphous polymer that will exhibit behavior characteristic of amorphous polymers. This, indeed, is so for polyethylene where a rubbery plateau and eventually a flow region are observed at temperatures above T_m.

The situation is somewhat similar in the case of polyvinyl chloride (PVC). PVC has a lower degree of crystallinity than polyethylene so that the viscoelastic response of PVC might be expected to approximate more closely that of an amorphous polymer than does polyethylene. Figure 3-3 indicates that this is indeed the case. Specifically, the modulus drop in PVC at T_g is considerably greater than that in polyethylene, although the rubbery plateau modulus is higher than that of amorphous polymers and the rubbery plateau extends to the melting point of the PVC crystallites, about 180°C. In a polymer of low crystallinity like PVC, the rubbery plateau can be well explained by the postulate that the crystallites act as both a filler and as crosslinks.

Once again, it should be pointed out that the exact shape of the modulus–temperature curve of a crystalline polymer depends on the thermal history of the sample, particularly on the rate of cooling from the melt and annealing treatment. Two crystalline polymers are mechanically "equivalent," for practical purposes, if they have the same values of T_g, T_m, chain length, percentage of crystallinity, and crystalline structure.

The incorporation of soluble diluents (plasticization) into polymers alters the viscoelastic properties to a greater or lesser extend depending on the exact polymer diluent system and the amount of diluent present. One of the most important polymers used in the plasticized state is polyvinyl chloride.

Figure 3-4 shows that PVC plasticized with 30-weight-percent dioctyl phthalate differs significantly from pure PVC in viscoelastic behavior. Although E_1 and E_2 appear to be essentially independent of the presence of diluents, other characteristic parameters show interesting effects. T_i, the inflection temperature, for example, is depressed, while the transition region is broadened. It was pointed out previously that rubbery flow commences in PVC when T_m is exceeded. Since T_m is depressed to 150°C due to the presence of diluent, the flow region begins at a lower temperature than for unplasticized PVC. If we regard diluent as part of the polymer system, then

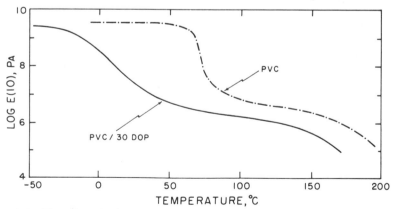

Figure 3-4. The effect of plasticization on a modulus–temperature curve. Reprinted from *Plasticization and Plasticizer Processes, Advances in Chemistry Series,* **48,** 1181 (1965) by permission of the American Chemical Society.

the occurrence of the flow region at lower temperatures might be interpreted as being due to the lowering of the number average molecular weight.

Since the plasticizer molecules are small, the argument used to rationalize the low glass transition temperature of the 10,000-molecular-weight polystyrene sample shown in Figure 3-2 seems applicable also to rationalize the decrease in the temperature of the glass-to-rubber transition in the plasticized system.

B. TIME–TEMPERATURE SUPERPOSITION

Remembering that the modulus is a function of time as well as temperature leads one to wonder about the feasibility of measuring the modulus as a function of time at constant temperature instead of, as we have done throughout the first part of this chapter, measuring the modulus as a function of temperature at constant time. In principle, the complete modulus-versus-time behavior of any polymer at any temperature can be measured. The experiment is straightforward, as described in Chapter 2, Section B. A number of convenient approaches are available. The results of such an experiment are shown on the left side of Figure 3-5 where tensile stress relaxation has been chosen as the experiment. Clearly only a small range of viscoelastic response manifests itself during an experimentally accessible time range of perhaps 10 seconds to 1 hour. Consider, for example, the experiment at temperature T_2. We observe that the modulus decays by a

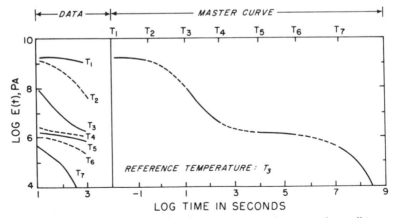

Figure 3-5. Preparation of a stress relaxation master curve from experimentally measured modulus–time curves at various temperatures.

factor of about 10 in this period of time, a decrease in modulus far short of the five orders of magnitude we observed in the modulus temperature experiments. If one were to carry out the experiment at a different temperature, different behavior would result; again this is shown on the left side of Figure 3-5. Clearly then, due to the broad time-dependence involved, it is not feasible to directly measure the complete behavior of the modulus as a function of time at constant temperature using methods such as those described in Chapter 2.

Experimentalists have found a shifting procedure that enables one to construct a "master curve" (complete modulus–time behavior at a constant temperature) in spite of this limitation. This technique is based on the principle of time–temperature correspondence.

Consider, now, the curve measured at T_1 and terminating at about 1000 seconds. The time–temperature correspondence principle states that there are two methods to use in order to determine the polymer's behavior at longer times. First, one may directly measure the response at longer times. This technique rapidly becomes prohibitively time-consuming because the change is so slow (Figure 3-5 is plotted on a log scale). Secondly, one may increase the temperature to T_2, for example, and again carry out the relaxation experiment on an experimentally accessible time-scale. Now one observes that shifting curve T_2 horizontally to the right will result in an exact superposition of the curves measured at T_1 and T_2 in the areas where the modulus values overlap and an extension of the curve measured at T_2 to modulus values lower than those measured at temperature T_1. The time–temperature correspondence principle states that this extension is identical to that which would be measured at long times at the temperature T_1. Thus, one effectively has a measure of the complete modulus–time behavior by applying the time–temperature correspondence principle to experimental measurements of polymer relaxations carried out on experimentally accessible time scales.

Mathematically these ideas may be expressed as

$$E(T_1, t) = E(T_2, t/a_T) \qquad (3\text{-}1)$$

where the effect of changing temperature is the same as applying a multiplicative factor to the time scale, i.e., an additive factor to the log time-scale.

Although this correspondence principle was proved to be enormously successful and generally applicable to amorphous polymers, Plazek[2] has shown that it is not quantitatively correct except in limited time-temperature ranges. It should be clear from the previous discussion that the principle cannot be expected to apply to multiphase or semicrystalline polymers.

When shifting experimental data, one additional correction is necessary. We have, by shifting horizontally, compensated for a change in the time scale brought about by changing temperature. There is also, however, an inherent change in the modulus brought about by a change in temperature. In Chapter 6 it is shown that the modulus of a rubbery network is directly proportional to T, the absolute temperature. Thus, in applying a reducing procedure to make a master curve for individual relaxation experiments, not only must one take into account the time-scale shift, one must also consider that there will be a slight vertical shift due to the temperature variation. Similarly, since the volume of a polymer is a function of temperature, and the modulus, being defined per unit cross-sectional area, will obviously vary with the amount of matter contained in unit volume, a corresponding correction must be made so that changing mass per unit volume as a function of temperature is accounted for. The density is obviously the parameter that must be used. These considerations lead one to write:[3]

$$\frac{E(T_1,t)}{\rho(T_1)T_1} = \frac{E(T_2,t/a_T)}{\rho(T_2)T_2} \qquad (3\text{-}2)$$

Division by the temperature corrects for the changes in modulus due to the inherent dependence of modulus on temperature while division by the density corrects for the changing number of chains per unit volume with temperature variation.

When constructing a master curve, one arbitrarily picks a reference temperature, T_0. The modulus at any time t, which one would observe at the temperature T_0 in terms of the experimentally observed modulus values at different temperatures T, is therefore given as

$$E(T_0,t) = \frac{\rho(T_0)T_0}{\rho(T)T} E(T,t/a_T) \qquad (3\text{-}3)$$

where a_T is to be discussed more fully. If one is considering compliance functions, equation (3-2) takes the form, using $J(t)$ as an example,

$$\rho(T_1)T_1 J(T_1,t) = \rho(T_2)T_2 J(T_2,t/a_T) \qquad (3\text{-}4)$$

and

$$J(T_0,t) = \frac{\rho(T)T}{\rho(T_0)T_0} J(T,t/a_T) \qquad (3\text{-}5)$$

C. MASTER CURVES

We may now consider the preparation of a master curve from the data in Figure 3-5. First, let us arbitrarily pick T_3 as the reference temperature. With a knowledge of the density at all the temperatures T_i, one applies the vertical correction factor stated in equation (3-3) to all the curves. For T_3 the factor $\rho(T_0)T_0/\rho(T)T$ is unity resulting in no shift. This must be so since T_3 has been chosen as the reference. At the other temperatures, however, the correction factor will not in general be unity. Let us now consider that these corrections have been made to the experimental curves on the left side of Figure 3-5. The curve at T_3 is reproduced on the right side of the figure. Next, the curve at T_2 is shifted to the left, giving rise to the dotted extension of the $E(T_3, t)$ curve. This process is repeated until the complete master curve is formed. Again this procedure is mathematically described by equation (3-3). The a_T's are functions of temperature and are known as the shift factors. The subscript T indicates that the shift factors are taken relative to some standard temperature. In our example, all of the curves at temperatures higher than T_3 are shifted to the right, those at T_1 and T_2 are shifted to the left. It should be clear that any temperature might have been chosen as the reference temperature. If T_4 had been chosen, for example, some of the shift factors would have been larger than 1.0—those at lower temperatures—while some of the shift factors would have been less than 1.0—those at temperatures higher than T_4.

D. THE WLF EQUATION

It is now common practice to reduce most relaxation or creep data to the temperature T_g; thus, the reference temperature is picked as the glass transition temperature measured by some slow technique such as dilatometry. The function $\log a_T$ for this choice of reference temperature is shown in Figure 3-6. Here again, at the reference temperature a_T is 1.0 and $\log a_T$ is 0. All amorphous polymers exhibit similar behavior on such a plot. The equation of this curve is:

$$\log a_T = \frac{-C_1(T-T_g)}{C_2+T-T_g} \tag{3-6}$$

The equation is called the WLF equation after its discoverers Williams, Landel, and Ferry.[4] The constants C_1 and C_2, originally thought to be universal constants, have been shown to vary rather slightly from polymer to polymer. A list of C_1 and C_2 for several of the most common polymers is presented in Table 3-2.

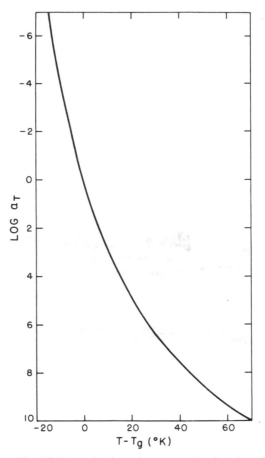

Figure 3-6. The WLF equation for polystyrene using data from Table 3-2.

Table 3-2. WLF Parameters[5]

Polymer	C_1	C_2	$T_g(°K)$
Polyisobutylene	16.6	104	202
Natural rubber (Hevea)	16.7	53.6	200
Polyurethane elastomer	15.6	32.6	238
Polystyrene[2]	14.5	50.4	373
Poly(ethyl methacrylate)	17.6	65.5	335
"Universal constants"	17.4	51.6	

The form of equation (3-6) is predicted in a straightforward way from rather simple theoretical considerations. First, however, we must realize that the shift factor a_T is not only meaningful in terms of moduli. In Chapter 7, Problem 7-8, the equation

$$\eta(T) = \int_0^\infty E(T,t)\,dt \tag{3-7}$$

is developed. $\eta(T)$ is the tensile viscosity of a polymer at the temperature T. Anticipating the results of Chapter 7, we use equation (3-7) in the argument that follows.

Consider two master curves at the different temperatures, T_g, which we will call *the* reference temperature, and T, any other temperature. We can write

$$E(T_g, t_{ref}) = E(T,t) \tag{3-8}$$

where we have constrained that every modulus value $E(T_g, t_{ref})$ has a corresponding modulus value of equal magnitude $E(T,t)$ on the master curve at temperature T. In writing this equation, we have implied a relationship between t_{ref} and t which, in the context of the above discussion, is realized to be

$$t_{ref} = \frac{t}{a_T} \tag{3-9}$$

where a_T is a function of T_g and T only. This is just another statement of time–temperature correspondence. (We are neglecting the vertical shifts for simplicity; their inclusion would substantially add to the complexity of this discussion but would not substantially affect the outcome.)

From equation (3-9) it is clear that

$$dt = a_T dt_{ref} \tag{3-10}$$

since a_T is not a function of time. Equation (3-7) becomes

$$\eta(T) = \int_0^\infty E(T,t)\,dt = \int_0^\infty E(T_g, t_{ref})\,dt \tag{3-11}$$

where the last expression results from substitution of equation (3-8) into equation (3-7). In spite of the similarity of notation, t_{ref} and t are different variables as is obvious from equation (3-9). Elimination of dt via equation

(3-10) yields

$$\eta(T) = \int_0^\infty E(T_g, t_{ref}) a_T \, dt_{ref} \tag{3-12}$$

where the limits of the integral are unchanged. Under this integration, however, a_T is a constant that can be factored out leaving an integral of the form of equation (3-7); thus

$$\eta(T) = a_T \eta(T_g)$$

$$a_T = \frac{\eta(T)}{\eta(T_g)} = \frac{\eta(T)}{\eta(T_{ref})} \tag{3-13}$$

where the latter expression results from the fact that we arbitrarily chose T_g as the reference temperature.

We may now return to the theoretical rationalization of the form of the WLF equation. The starting point is the semiempirical Doolittle equation for the viscosity of a liquid[6]

$$\ln \eta = \ln A + B \left(\frac{V - V_f}{V_f} \right) \tag{3-14}$$

which gives an expression for the viscosity of a system in terms of two constants A and B. We will assume this viscosity is the tensile viscosity. V is the total volume of the system while V_f is the free volume available to the system (a qualitative rather than quantitative view of free volume is sufficient for this discussion). The interpretation of equation (3-14) is that viscosity is intimately connected with mobility which, in turn, is closely related to free volume. As the free volume increases, the viscosity rapidly decreases. This equation has been found to express the viscosity dependence of simple liquids to a high degree of accuracy. Rearrangement of equation (3-14) gives

$$\ln \eta = \ln A + B \left(\frac{1}{f} - 1 \right) \tag{3-15}$$

where f is the fractional free volume V_f/V. It is now assumed that above the glass transition temperature, the fractional free volume increases linearly with temperature, that is:

$$f = f_g + \alpha_f (T - T_g) \tag{3-16}$$

where f is the fractional free volume at T, any temperature above T_g, f_g is the fractional free volume at T_g, and α_f is the coefficient of thermal expansion of the fractional free volume above T_g. In terms of equation (3-16), the Doolittle equation becomes

$$\ln \eta(T) = \ln A + B\left(\frac{1}{f_g + \alpha_f(T - T_g)} - 1\right) \qquad \text{at} \quad T > T_g$$

$$\ln \eta(T_g) = \ln A + B\left(\frac{1}{f_g} - 1\right) \qquad \text{at } T_g \tag{3-17}$$

Subtraction yields

$$\ln \frac{\eta(T)}{\eta(T_g)} = B\left(\frac{1}{f_g + \alpha_f(T - T_g)} - \frac{1}{f_g}\right) \tag{3-18}$$

which simplifies to

$$\log \frac{\eta(T)}{\eta(T_g)} = \log a_T = -\frac{B}{2.303 f_g}\left(\frac{T - T_g}{(f_g/\alpha_f) + T - T_g}\right) \tag{3-19}$$

a form identical to the WLF equation where C_1 is identified with $B/2.303 f_g$ and C_2 with f_g/α_f. Thus the form of the WLF equation is consistent with the accurate empirical Doolittle equation and the assumption of a linear expansion of free volume above T_g.

The complete viscoelastic response of any polymer under any experimental conditions may be obtained from a knowledge of any two of the following three functions: the master curve at any temperature, the modulus–temperature curve at any time, and the shift factors relative to some reference temperature. For example, suppose we are given the constants C_1 and C_2 for a polymer whose master curve is known. For simplicity, we can assume that the master curve is at the same reference temperature as that in the WLF equation, perhaps T_g. Suppose it is desired to calculate the 10-second modulus-versus-temperature curve for this polymer. The 10-second modulus at T_g is read directly from the master curve. Now, however, the master curve can be shifted to exhibit the behavior of the polymer at some other temperature. The amount of shift on a log scale, $\log a_T$, is given by equation (3-6) where the C_1 and C_2 used are those given. Applying this horizontal shift, with the slight additional vertical correction, if significant, allows one to "predict" the 10-second modulus, at this new temperature from the shifted curve. This procedure is repeated until the entire modulus–time curve is generated (Figure 3-7).

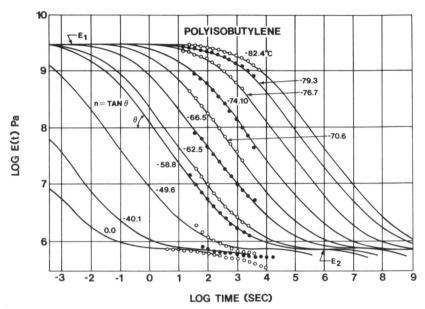

Figure 3-7. Plot of a stress relaxation master curve at various temperatures. Reprinted from A. V. Tobolsky and H. F. Mark, Eds., *Polymer Science and Materials*, p. 223, by permission of John Wiley & Sons, Inc.

Having thus generated the manifold of curves shown in Figure 3-7, it is possible to view the dependence of the modulus–temperature plots on the arbitrary choice of time. This is done merely by picking points off intersections of the master curves with vertical lines drawn from the point of interest on the time scale. The results of such treatment of the data[7] are shown in Figure 3-8. It is apparent that the longer the constant time of measurement, the sharper the resulting curves. In fact, if it is assumed that the ideas embodied in the WLF equation are applicable at temperatures considerably below T_g, it can be shown that an experiment of the type depicted in Figure 3-8 carried out infinitely slowly would result in a true second-order thermodynamic transition, that is, a discontinuity in the modulus, at a temperature about 50°C below T_g.

E.   MOLECULAR INTERPRETATION OF VISCOELASTIC RESPONSE

We may now briefly discuss the molecular interpretation of the behavior exhibited by polymers that are held at constant temperature and studied as a function of time.

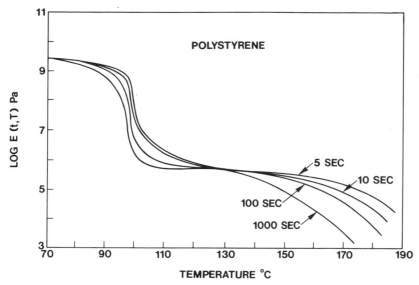

Figure 3-8. Plot of modulus-versus-temperature curves for various values of the constant time. Reprinted from A. V. Tobolsky *Properties and Structure of Polymers*, page 74, by permission of John Wiley & Sons, Inc.

In the very short time-ranges, the molecules, not having sufficient time to reorient substantially, probably react to a stress by distorting intermolecular distances. These distortions, being of a rather high energy, result in a high modulus. At longer times, however, reorientation of chain segments is possible and occurs rather extensively. The chains reorient in order to relieve local strains imposed by the stress. Thus, severely pinched interactions relax to lower energy conformations, resulting in a modulus decrease. This is the glass transition process. At still longer times, the chains are, on the average, stretched. The average "end-to-end" length (Chapter 4) of the entangled chains has increased. This elongation is maintained due to long-range interactions that prevent long-range motions. Such a network of interacting flexible chains is a rubber. In the longest time range of interest, the chains can move past one another, resulting in complete relaxation and a modulus indicative of a liquid. Now the end-to-end distance returns to its equilibrium value.

In this discussion, it is impossible to describe quantitatively the time ranges that give rise to each type of behavior, since the temperature variable causes all of these ranges to be relative.

According to these ideas, a plastic would have the modulus of a rubber on a time scale of perhaps thousands of years while a rubber might behave like a plastic on a nanosecond time scale.

Direct evidence of this type of behavior is provided by the change in shape of the glass windows in some of the oldest cathedrals in Europe. On the time scale of many centuries, the glass is a rather viscous liquid; the bottoms of these windows are much thicker than the tops due to the response of this viscous liquid to the force of gravity. On the time scale of days, however, the material is hard, having withstood occasional high forces due to wind or pounding rain without appreciable response.

PROBLEMS

1. An apparent activation energy for viscoelastic relaxation may be obtained as the slope (multiplied by R) of a plot of $\ln a_T$ as a function of $1/T$. Such a plot yields a curved line and thus a temperature-dependent activation energy.[8]

(a) Obtain an analytical expression for the activation energy from the WLF equation. Using the "universal" constants C_1 and C_2, what is the value of the activation energy at T_g if (1) $T_g = 200°K$ (2) $T_g = 400°K$?

(b) Show that the energy becomes independent of temperature for $T \gg T_g$ and approaches the value of 4.1 kcal for all materials. Would you expect the activation energy for real polymers to obey this prediction? Why or why not?

2. The WLF equation may be written using any convenient temperature as reference temperature. The form of the equation remains the same but the values of the constants C_1 and C_2 change. Using the "universal" values for C_1 and C_2 at T_g, calculate C_1 and C_2 for a reference temperature of $T_g + 50°C$.

3. Tabulated below are the results of stress relaxation measurements made on bis-phenol-A-polycarbonate.[9] Construct the master curve and calculate C_1 and C_2, the WLF equation parameters, for a reference temperature of 150.8°C.

Stress Relaxation—Polycarbonate $M_\omega = 40,000$
(After Mercier, J. P. et al., J. Appl. Polym. Sci., 9, 447 [1965])

$T(°C)$	$\log t$	$\log E(t)$	$T(°C)$	$\log t$	$\log E(t)$
130	2.00	10.25	141	2.00	10.19
	2.25	10.25		2.25	10.19
	2.50	10.25		2.50	10.16
	2.75	10.25		2.75	10.13
	3.00	10.22		3.00	10.09
	3.25	10.19		3.25	10.03
	3.50	10.16		3.50	9.94
	3.75	10.13		3.75	9.82
	4.00	10.06		4.00	9.66
	4.25	10.03		4.25	9.50

Stress Relaxation—Polycarbonate $M_\omega = 40,000$ *(Continued)*

$T(°C)$	$\log t$	$\log E(t)$	$T(°C)$	$\log t$	$\log E(t)$
142	2.00	10.13	156	2.50	7.50
	2.25	10.09		2.75	7.41
	2.50	10.03		3.00	7.34
	2.75	9.97		3.25	7.25
	3.00	9.88		3.50	7.19
	3.25	9.67		3.75	7.06
	3.50	9.63		4.00	6.94
	3.75	9.47		4.25	6.78
	4.00	9.22		4.50	6.59
144.9	2.25	9.66		4.75	6.38
	2.50	9.50		5.00	6.13
	2.75	9.28	159	2.25	7.34
	3.00	9.03		2.50	7.25
	3.25	8.88		2.75	7.16
	3.50	8.66		3.00	7.03
145.6	2.75	9.03		3.25	6.94
	3.00	8.72		3.50	6.78
	3.25	8.50		3.75	6.63
	3.50	8.25		4.00	6.41
	3.75	8.09		4.25	6.13
	4.00	7.94		4.50	5.85
146	2.50	8.72		4.75	5.53
	2.75	8.47	161.5	2.25	7.06
	3.00	8.25		2.50	6.97
	3.25	8.06		2.75	6.85
	3.50	7.94		3.00	6.64
	3.75	7.82		3.25	6.50
	4.00	7.67		3.50	6.31
150.8	2.00	7.97		3.75	6.06
	2.25	7.85		4.00	5.78
	2.50	7.78		4.25	5.47
	2.75	7.72	167	2.25	6.72
	3.00	7.69		2.50	6.50
	3.25	7.64		2.75	6.25
	3.50	7.59		3.00	5.97
	3.75	7.55		3.25	5.63
	4.00	7.50		3.50	5.22
	4.25	7.44	171	2.25	6.00
	4.50	7.34		2.50	5.72
	4.75	7.25		2.75	5.41
156	2.00	7.59		3.00	5.03
	2.25	7.55		3.25	4.63

4. Determine the 10-second modulus-versus-temperature curve for a material whose tensile stress relaxation modulus is given as

$$E(t, T_g) = (3.0 \times 10^9 e^{-t/\text{sec}} + 5.0 \times 10^5 e^{-t/10^4 \text{sec}}) Pa$$

Use the "universal" values for the constants in the WLF equation.

5. Derive an expression for $\log a_T$ for a polymer whose shear stress relaxation modulus is given as

$$G(t, T) = \sum_{i=1}^{N} G_i e^{-t/\tau_i(T)}$$

where each G_i is independent of temperature and each τ_i is of the Arrhenius form:

$$\tau_i = A_i e^{H/RT}$$

A_i is constant independent of T. Note also that H is constant for all relaxation times.

6. The Vogel equation,

$$\log a_T = \frac{1}{\alpha(T - T_0)} - \beta$$

is often used to treat the temperature-dependence of viscoelastic properties. Show that the Vogel equation is identical to the WLF equation by expressing α, β and T_0 in terms of C_1, C_2, and T_g.

REFERENCES

1. A. V. Tobolsky and H. F. Mark, *Polymer Science and Materials*, Wiley, New York, 1971.
2. D. J. Plazek, *J. Phys. Chem.*, **69**, 3480 (1965).
3. A. V. Tobolsky and J. R. McLoughlin, *J. Polym. Sci.*, **8**, 543 (1952).
4. M. L. Williams, R. F. Landel, and J. D. Ferry, *J. Amer. Chem. Soc.*, **77**, 3701 (1955).
5. J. D. Ferry, *Viscoelastic Properties of Polymers*, 3rd ed., Wiley, New York, 1980.
6. A. K. Doolittle and D. B. Doolittle, *J. Appl. Phys.*, **31**, 1164 (1959).
7. A. V. Tobolsky, *Properties and Structure of Polymers*, Wiley, New York, 1960.
8. E. Catsiff and A. V. Tobolsky, *J. Colloid Sci.*, **10**, 375 (1955).
9. J. P. Mercier et al., *J. Appl. Polym. Sci.*, **9**, 447 (1965).

4

Transitions and Relaxation in Amorphous Polymers

Amorphous polymers, as we have seen in Chapter 3, undergo a transition from glassy behavior to rubbery behavior. This transition is quite sharp, usually occurring over a temperature range of a few degrees. The decrease in modulus of about 3 orders of magnitude that accompanies this transition makes it clear that it is the single most important parameter characterizing the mechanical behavior of amorphous polymers.

In this chapter, the glass transition phenomenon is treated in moderate detail, including phenomenological aspects, molecular theories, and the effect of molecular structure. In addition, relaxation occurring in the glassy state below the glass transition temperature, (T_g), is discussed.

A. PHENOMENOLOGY OF THE GLASS TRANSITION

The classic method for the experimental determination of the glass transition temperature is dilatometry. Thus, as briefly mentioned in Chapter 1, the temperature dependence upon cooling of the specific volume is determined by a suitable technique, and the temperature at the change in slope is taken as T_g. Such a plot is indicated in Figure 4-1, where it is shown that the T_g is actually determined as the intersection of the straight-line portions

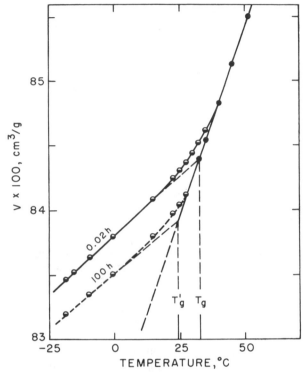

Figure 4-1. Specific volume versus temperature for a polyvinylacetate sample quick-quenched from well above T_g to the experimental temperatures. Volumes measured 0.02 hours and 100 hours after quenching as indicated. Filled points represent equilibrium behavior. After A. J. Kovacs, *J. Polym. Sci.*, **30**, 131 (1958).

of the curve above and below T_g. The slope is, of course, related to the cubical coefficient of thermal expansion, α, which exhibits a discontinuity at T_g. As is apparent from Figures 4-1 and 4-2, the glass transition is a rate-dependent phenomenon.

Dilatometric methods are simple in principle but complex in practice. The displacement method is perhaps the most commonly used. The thermal expansion coefficient is determined by measuring the amount of confining fluid displaced by the polymer. The success of the method depends on the selection of a confining fluid that is not absorbed by the polymer and undergoes no phase transitions in the temperature range of interest. Mercury is often used although it possesses obvious temperature limitations.

In addition to dilatometry, calorimetry has been extensively employed, a discontinuity or peak in heat capacity being observed at T_g, depending on

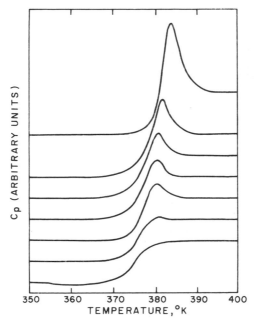

Figure 4-2. Heat capacity versus temperature measured at constant heating rate (9.0×10^{-2} °C/sec) for polystyrene. Linear cooling rates were, from top to bottom: 1.4×10^{-4}, 3.2×10^{-3}, 1.8×10^{-2}, 4.13×10^{-2}, 8.7×10^{-2} and .5°C/sec. Cooling started well above T_g and successive curves have shifted ordinates. After B. Wunderlich, D. M. Bodily and M. H. Kaplan, J. Appl. Phys., **35**, 95 (1964).

thermal history and heating rate (Figure 4-2). The recent developments of differential thermal analysis and differential scanning calorimetry[1] have enhanced the importance of thermodynamic measurements in the determination of T_g.

It has been repeatedly emphasized throughout this book that the glass transition in amorphous polymers is accompanied by profound changes in their viscoelastic response. Thus the stress relaxation modulus commonly decreases by about three orders of magnitude in the vicinity of the T_g determined by calorimetry or dilatometry, and the creep compliance increases by about three orders of magnitude. In addition, under dynamic experimental conditions, the storage moduli and compliances behave in a similar manner to the corresponding static quantities. The loss moduli and compliances, on the other hand, exhibit maxima in the glass transition region, as does $\tan \delta$. Figure 4-3 summarizes these results for a styrene–butadiene copolymer. The data in Figure 4-3 were collected at a constant frequency of 1 Hz over the temperature range indicated. (The logarithmic

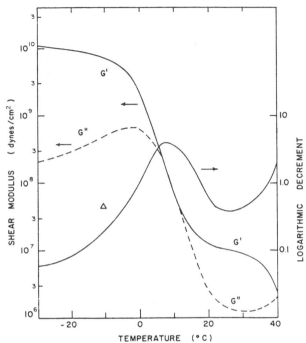

Figure 4-3. Temperature dependencies of G', G'', and the logarithmic decrement, Δ, for a styrene butadiene copolymer. After L. E. Nielsen, *Mechanical Properties of Polymers*, Reinhold, New York, 1962.

decrement, Δ, is plotted in Figure 4-3 rather than $\tan \delta$, but is merely related to $\tan \delta$ as $\Delta \approx \pi \tan \delta$.) It can be seen from Figure 4-3 that the maxima in $\tan \delta$ and G'' occur at different temperatures, the G'' maximum being lower. Frequently, the $\tan \delta$ or G'' peak temperatures are taken as the T_g but it is clear that they will not be identical and that in general they will both yield higher values than those obtained by dilatometric or thermodynamic methods, owing to differences in the time scale. Figure 4-4 gives the temperature dependence of E' and $\tan \delta$ for polyvinyl chloride at various frequencies. The shift of the $\tan \delta$ peak maximum to higher temperatures with increasing frequency occurs in accordance with the WLF equation and is again an illustration of the rate dependence of the glass transition phenomenon.

One of the oldest methods for the determination of T_g involves the temperature dependence of viscosity. Figure 4-5 is a plot of the variation of viscosity with temperature for polymethyl methacrylate. Various authors have argued that the glass transition temperature represents an isoviscous state with viscosity 10^{13} poises. Such a high viscosity is not easily measur-

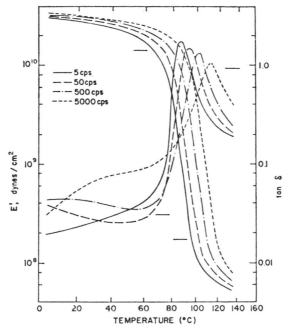

Figure 4-4. Temperature dependencies of E' and $\tan \delta$ for polyvinyl chloride at various frequencies. After G. W. Becker, *Kolloid-Z*, **140**, 1 (1955). Reprinted by permission of Dr. Dietrich Steinkopff Verlag.

able by normal flow techniques and, in fact, its determination is usually accomplished from creep or stress relaxation measurements by the techniques of Chapter 7.

Many profound physical changes occur at T_g in addition to those already mentioned. The refractive index undergoes an abrupt change at T_g. In the case of polar polymers, the dielectric loss tangent passes through a maximum in the vicinity of T_g, as does the imaginary part of the complex dielectric constant (Chapter 8, Section E). Because of the increased mobility at T_g, the line width in a nuclear magnetic resonance experiment undergoes an abrupt narrowing at this temperature. All of these effects, and many others, have been used as measures of T_g, and they all arise from the fundamental changes in molecular dynamics which occur in the vicinity of T_g. It should be apparent from the discussion that each experimental method will yield a different value for T_g and that even the same method will yield different results depending on the time scale. Thus, in order to compare T_g values, both the method of measurement and the rate of measurement should be specified.

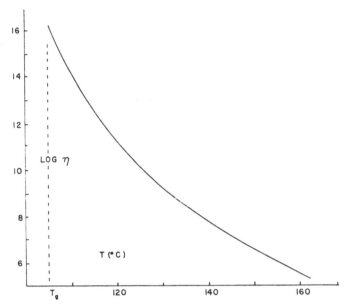

Figure 4-5. Temperature dependence of the viscosity for polymethyl methacrylate with $M = 63,000$. Reprinted from F. Bueche, *Physical Properties of Polymers*, p. 101, by permission of John Wiley & Sons, Inc.

B. THEORIES OF THE GLASS TRANSITION

1. Free Volume Theory

In Section D of Chapter 3 the WLF equation was derived on the basis of free volume concepts. In particular, we may write

$$\log a_T = \frac{B}{2.303}\left(\frac{1}{f_T} - \frac{1}{f_{T_0}}\right) \tag{4-1}$$

where f_T and f_{T_0} are the fractional free volumes at temperatures T and T_0, respectively, and B, a parameter in the Doolittle equation, is experimentally found to be close to unity.

Unfortunately, the concept of free volume is difficult to define in a precise manner. In an approximate way we can represent the segments of a polymer chain by rigid bodies and the free volume as the holes present between these segments as a result of packing requirements. Presumably the free volume reaches a constant value at T_g that is too small to allow the large-scale conformational rearrangements of the chain backbones associated with T_g to occur. Above T_g, on the other hand, free volume increases and becomes sufficiently large to allow such motions to occur. These ideas are embodied in equation (3-16), which we requote here.

$$f = f_g + \alpha_f(T - T_g) \qquad T \geqslant T_g$$
$$f = f_g \qquad\qquad\quad T < T_g \qquad\qquad (3\text{-}16)$$

where the fractional free volume reaches a constant value, f_g, at T_g and increases linearly above T_g with the coefficient of expansion α_f.

Following Chapter 3, substitution of equation (3-16) into equation (4-1) with T_g as the reference temperature yields the WLF equation where the constants C_1 and C_2 are given by

$$C_1 = \frac{B}{2.303 f_g} \qquad C_2 = \frac{f_g}{\alpha_f} \qquad\qquad (4\text{-}2)$$

Knowledge of the numerical values of C_1 and C_2 thus leads to the parameters f_g and α_f through equation (4-2), since B is taken as unity. The constants C_1 and C_2 were originally taken to be universal for all amorphous polymers with $C_1 = 17.44$ and $C_2 = 51.6$. It was later found that C_1 values were indeed approximately constant for all systems but that C_2 varied quite widely. This is illustrated by Table 3-2. Equation (4-2) indicates that this result means that f_g is approximately constant at a value of 0.025, but that α_f varies from one amorphous polymer to another. It is clear that the WLF equation predicts that T_g represents an iso-free volume state. While this concept is not strictly true it is nevertheless of wide utility, as we shall see in the following discussion.

It is possible to identify α_f with $\Delta\alpha = \alpha_r - \alpha_g$, that is, the difference between the expansion coefficients above and below T_g. If this is done, good agreement is obtained for many polymers between experimentally determined values of $\Delta\alpha$ and α_f obtained from C_2 values through equation (4-2). Inasmuch as there is no *a priori* method of defining α_f, the above agreement is perhaps the best justification for the identification of α_f with $\Delta\alpha$.

The free volume approach embodied in equation (3-16) may be easily generalized to include the effect of pressure. Intuitively, pressure may be thought of as having the effect of "squeezing out" free volume and thus raising the glass transition temperature; such has actually been found to be the case. The magnitude of the effect is predictable as follows. Assume the polymer is taken from state 1 characterized by P_1 and T_{g_1} to state 2 characterized by T_{g_2} and P_2. We may write in general for any change from state 1 to state 2,

$$f_2 = f_1 + \alpha_f(T_2 - T_1) - \beta_f(P_2 - P_1) \tag{4-3}$$

In equation (4-3) β_f refers to the isothermal compressibility of free volume, assumed independent of pressure over the range of interest. If the polymer remains at T_g in going from state 1 to state 2, f_2 must also remain equal to f_1 since T_g is an iso-free volume state. Thus, under these conditions,

$$\alpha_f(T_{g_2} - T_{g_1}) = \beta_f(P_2 - P_1) \tag{4-4}$$

For small changes, equation (4-4) becomes

$$\frac{dT_g}{dP} = \frac{\beta_f}{\alpha_f} \tag{4-5}$$

Let $\beta_f = \Delta\beta$, the compressibility above and below T_g, and $\alpha_f = \Delta\alpha$. Then

$$\frac{dT_g}{dP} = \frac{\Delta\beta}{\Delta\alpha} \tag{4-6}$$

Although precise pressure measurements are difficult and not very much data have been reported, amorphous polymers at least approximately obey equation (4-6). (See Table 4-1.)

The free-volume theory finds ready application in predicting the effect on T_g of diluents and molecular weight, among others. As an example, we shall derive an expression for the dependence of T_g on plasticizer concentration. The total fractional free volume of a polymer–diluent system may

Table 4-1. Pressure Dependence of T_g

Polymer	$TV\Delta_\alpha/\Delta C_p$ (°C/atm)	dT_g/dP (°C/atm)	$\Delta\beta/\Delta\alpha$ (°C/atm)
Polyvinyl acetate	0.025[a]	0.02[a]	0.05[a]
Polystyrene	—	0.036[b]	0.10[b]
Natural Rubber	0.020[a]	0.024[c]	0.024[c]
Polymethyl methacrylate	—	0.023[d]	0.065[e]

[a] J. M. O'Reilly, *J. Polym. Sci.*, **57**, 429 (1962).
[b] S. Matsuoka and B. Maxwell, *J. Polym. Sci.*, **32**, 131 (1958).
[c] J. E. McKinney, H. V. Belcher, and R. S. Marvin, *Trans. Soc. Rheo.*, **4**, 347 (1960).
[d] N. Shishkin, *Sov. Phys.—Solid State*, **2**, 322 (1960).
[e] G. Allen, D. Sims, and G. J. Wilson, *Polymer*, **2**, 375 (1961).

be written as

$$f = 0.025 + \alpha_p (T - T_{g_p}) V_p + \alpha_d (T - T_{g_d}) V_d \qquad (4\text{-}7)$$

In equation (4-7) the subscripts p and d refer to polymer and diluent, respectively, and V is the volume fraction. At the T_g of the polymer–diluent mixture, f becomes 0.025. Using this fact and substituting $1 - V_p$ for V_d, we have

$$T_g = \frac{\alpha_p V_p T_{g_p} + \alpha_d (1 - V_p) T_{g_d}}{\alpha_p V_p + \alpha_d (1 - V_p)} \qquad (4\text{-}8)$$

The dependence of T_g on V_p as predicted from equation (4-8) is plotted in Figure 4-6 for the system polymethyl methacrylate–diethyl phthalate, together with some experimental results. The calculated curve is seen to agree well with the experimental data.

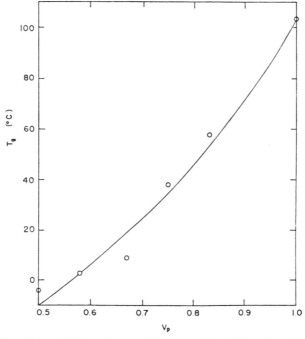

Figure 4-6. Dependence of T_g on V_p for mixtures of polymethyl methacrylate with diethyl phthalate. Comparison of experimental results with equation (4-8). Parameters used are $\alpha_d = 10 \times 10^{-4}\,^\circ\text{C}^{-1}$, $T_{gd} = -65\,^\circ\text{C}$. After M. Gordon and J. S. Taylor, *J. Appl. Chem.*, **2**, 493 (1952), by permission of the Society of Chemical Industry.

2. Thermodynamic Theory

Equilibrium phase transitions are well treated using the classical thermodynamic approach. In particular, the phenomena of melting and vaporization are characterized by the equality of the Gibbs free energy of the two phases at the transition temperature.

$$F_1 = F_2 \tag{4-9}$$

Also

$$dF_1 = dF_2 \tag{4-10}$$

However, the volumes and entropies of the two phases are not equal.

$$S_1 \neq S_2$$
$$V_1 \neq V_2 \tag{4-11}$$

Ehrenfest[2] refers to the type of phase transition described above as a first-order transition because there are discontinuities in the first partial derivatives of the Gibbs free energy at the transition point.

For example,

$$dF = -S\,dT + V\,dP \tag{4-12}$$

and we have

$$S = -\left(\frac{\partial F}{\partial T}\right)_P$$
$$V = \left(\frac{\partial F}{\partial P}\right)_T \tag{4-13}$$

This notion is easily generalized to higher-order transitions. Thus a second-order transition is described as one in which the second partial derivatives of the Gibbs free energy show discontinuities at the transition point, a third-order transition involves discontinuities in the third partial derivatives, and so on.

In particular, for a second-order transition,

$$-\left(\frac{\partial^2 F}{\partial T^2}\right)_P = \left(\frac{\partial S}{\partial T}\right)_P = \frac{C_p}{T}$$

$$\left(\frac{\partial^2 F}{\partial P^2}\right)_T = \left(\frac{\partial V}{\partial P}\right)_T = -\beta V \tag{4-14}$$

$$\left(\frac{\partial}{\partial T}\left(\frac{\partial F}{\partial P}\right)_T\right)_P = \left(\frac{\partial V}{\partial T}\right)_P = \alpha V$$

So, at the second-order transition temperature, the following discontinuities are observed:

$$\Delta C_p = C_{p_2} - C_{p_1}$$

$$\Delta \beta = \beta_2 - \beta_1 \qquad (4\text{-}15)$$

$$\Delta \alpha = \alpha_2 - \alpha_1$$

It is just these discontinuities that often occur in experiments involving the glass transition, and, because of this, T_g is often referred to as the second-order transition point. It is clear, however, that the observed T_g is a rate-dependent phenomenon. The thermodynamic analysis given above refers only to an equilibrium process and thus cannot be directly applied to the experimental T_g. The kinetic nature of the observed T_g does not, however, preclude the existence of a true second-order transition temperature. Thus, in a dilatometric experiment, when the polymer is cooled from the rubbery or liquid states, volume contractions take place involving conformational rearrangements. At temperatures far above T_g thermal equilibrium is maintained during the cooling process, but as the temperature is lowered a point is eventually reached at which the rate of volume contraction becomes comparable with the rate of cooling. Below this temperature, volume relaxation and hence conformational rearrangements are not possible during the time scale of the experiment, and the discontinuities ΔC_p, $\Delta \beta$, and $\Delta \alpha$ are observed. Such an analysis indicates that an infinitely slow cooling rate would be required to observe the true thermodynamic second-order transition temperature, if it exists at all.

Kauzmann[3] noted that if the entropies of simple glass-forming liquids were extrapolated to low temperatures, they would go to zero long before the absolute zero of temperature was reached. Such a result, requiring negative entropies, is clearly meaningless in a physical sense. Kauzmann resolved this paradox by suggesting that the glassy state is not an equilibrium state and that the glasses would pass over into crystalline solids at equilibrium before the temperature of zero-extrapolated entropy could be achieved. Such an explanation denies the existence of a second-order thermodynamic transition. However, many polymeric substances have never been obtained in a crystalline state and it is difficult to imagine the possibility of crystallizing a material such as atactic polystyrene, for example. It was subsequently suggested by Gibbs and DiMarzio[4] that the conformational entropy at *equilibrium* would indeed attain a value of zero at a finite temperature, defined as T_2, undergoing no further change between T_2 and absolute zero. This idea is schematically illustrated in Figure 4-7 in which the dotted line represents Kauzmann's extrapolation.

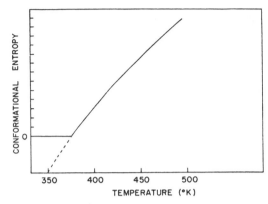

Figure 4-7. Schematic plot of conformational entropy versus temperature for a glass-forming substance. The temperature at which the entropy attains a value of 0 is the T_2 of Gibbs and DiMarzio. The dotted line is the original extrapolation of Kauzmann. After J. H. Gibbs in *Modern Aspects of the Vitreous State*, Vol. 1. Reprinted by permission of Butterworth and Co.

The T_2 of Gibbs and DiMarzio is a true second-order transition temperature. In this view the observed T_g is the kinetic reflection of the underlying thermodynamic phenomenon that would occur at T_2.

Gibbs and DiMarzio suggest that at high temperatures a very great number of conformations are accessible to each chain. They assume that for each chain segment there is one definite lowest energy conformation, so that at equilibrium at low temperatures there is just *one* conformation available to the entire chain. Thus, as the chain is cooled from high to low temperatures, fewer and fewer high-energy conformations become accessible until at T_2 only the one lowest energy state is allowed.

In reality, however, there are so many states accessible at high temperatures that the reorientation from one state to any other particular state may take quite some time. In fact, as we cool the sample and intrinsically slow down molecular movement, it should take longer and longer to get to any particular conformation. To insure getting all the chains into the lowest energy conformation, the experiment would have to be carried out infinitely slowly.

The temperature at which the second-order transition would occur according to these ideas may be calculated using the WLF equation (3-6). To shift an experiment from a finite time-scale to an infinite time-scale would take a value of $\log a_T$ approaching infinity. Clearly this will be the case if the denominator of the right side of equation (3-6) goes to zero while the numerator remains finite.

$$C_2 + T_2 - T_g = 0 \qquad (4\text{-}16)$$

Since C_2 is relatively constant for most polymers, one derives the relation that

$$T_2 = T_g - C_2 \approx T_g - 52 \qquad (4\text{-}17)$$

Thus the second-order phase change would be observed about 50°C below T_g for an experiment carried out infinitely slowly.

The Gibbs–DiMarzio model is characterized by two parameters. These are the hole-formation energy, u_0, and the flex energy, ε. The parameter u_0 is a reflection of intermolecular energy contributions and ε is a reflection of intramolecular energy contributions. Specifically, for hydrocarbon chains such as polymethylene, u_0 is the energy of interaction, or the "van der Waals bond" energy between a pair of chemically nonbonded but nearest neighboring segments in the lattice. In order to define ε, we note that if Z is the number of primary valences of each backbone chain atom, there are $(Z-1)$ possible orientations of a bond i with respect to the coordinate system formed by the bonds $(i-1)$ and $(i-2)$ of the same molecule. In the case of a hydrocarbon chain, $Z = 4$. It is assumed that an energy ε_1 is associated with one of the three possible orientations and an energy ε_2 is associated with the other two, with $\varepsilon_2 > \varepsilon_1$. Then ε is defined as the difference between ε_2 and ε_1. It should be emphasized that ε, a thermodynamic quantity, has nothing to say about the magnitude of potential energy barriers to rotation about chain backbone bonds. Rather, ε refers only to the energy difference between low-lying rotational isomeric states. According to the theory, it is found that u_0 is directly proportional to T_2 and that ε/kT_2 is a constant for all polymers. Recasting this ratio in terms of T_g rather than T_2, it can be shown that ε/kT_g possesses the "universal" value of 2.26 for all amorphous polymers.[5]

We can demonstrate the utility of the Gibbs–DiMarzio theory by applying it to the calculation of the T_g of a random copolymer. Let us assume that the flex energy of a random copolymer composed of A and B monomer units is given by

$$\varepsilon = X_A \varepsilon_A + X_B \varepsilon_B \qquad (4\text{-}18)$$

where X_A and X_B are the mole fractions of monomer units A and B, respectively. Substituting the "universal" value of $\varepsilon = 2.26 kT_g$ in equation (4-18) immediately leads to

$$T_g = X_A T_{g_A} + X_B T_{g_B} \qquad (4\text{-}19)$$

which is identical in form to an empirical equation of Wood[6] found to have wide applicability to random copolymers.

The Gibbs–DiMarzio approach, as has been stated, rests on the existence of a true second-order thermodynamic transition at a temperature T_2 below the observed T_g. The pressure dependence of T_2 can thus be obtained by the methods of equilibrium thermodynamics. For a first-order phase transition, the pressure dependence of the transition temperature is given by the Clapeyron equation

$$\frac{dT}{dP} = \frac{\Delta V}{\Delta S} \tag{4-20}$$

Straightforward application of the Clapeyron equation to a second-order transition is not possible because both ΔV and ΔS are 0, and thus dT/dP is indeterminate. However, we may invoke L'Hopital's rule and differentiate both numerator and denominator of the right side of equation (4-20) independently to obtain the limiting behavior. This may be done with respect to either temperature or pressure. Thus, recalling equation (4-14) and differentiating equation (4-20) with respect to temperature:

$$\frac{dT_2}{dP} = \frac{\partial \Delta V/\partial T}{\partial \Delta S/\partial T} = \frac{V \Delta \alpha T_g}{\Delta C_p} \tag{4-21}$$

Differentiating with respect to pressure, there results

$$\frac{dT_2}{dP} = \frac{\partial \Delta V/\partial P}{\partial \Delta S/\partial P} = \frac{\Delta \beta}{\Delta \alpha} \tag{4-22}$$

Thus the pressure dependence of T_2 and hence T_g is given by either equation (4-21) or equation (4-22). Equation (4-22) is, of course, identical to the free-volume result, equation (4-6). A comparison of these quantities has been given in Table 4-1.

3. Kinetic Theories

Several kinetic theories have been particularly successful in explaining various features of the glass transition phenomenon. Here the emphasis is on separating fundamental molecular aspects of the transition from factors introduced by the particular way in which an experiment is carried out. Although such theories have become moderately complicated in the past several years, by concentrating on single-ordering parameter models we will attempt to convey the general framework of these theories without tackling the conceptual and mathematical complications that are necessary to realize quantitative agreement between theory and experiment.

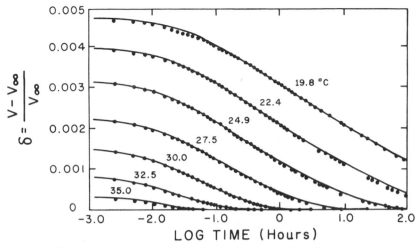

Figure 4-8. Contraction isotherms for glucose, a low-molecular-weight glass-forming substance. Samples were quick-quenched from equilibrium at 40°C to the temperatures indicated. Solid lines were calculated using equations (4-23) and (4-25) with $T_g = 35°C$, $f_g = 0.025$, $\alpha_f = 3.6 \times 10^{-4}°C^{-1}$, and $\tau(T_g, \delta = 0) = 0.015$ hours. After A. J. Kovacs, *Adv. Polym. Sci.*, 3, 394 (1964).

For the sake of this discussion, the data to be analyzed are of a slightly different sort from those considered previously. In Figure 4-8, we have plotted several volume-contraction isotherms for glucose. First of all, it should be noted that glucose is not a high-molecular-weight polymer, but rather a low-molecular-weight material that is easily vitrified. In fact, many aspects of the glass transition phenomenon are virtually independent of molecular weight. Thus, the behavior exhibited by glucose in its glass transition region is essentially the same as that of any high-molecular-weight amorphous polymer. The experiments represented in this figure involve annealing the sample at 40°C (a temperature above the usually quoted glass transition temperature, $T_g = 35°C$) until equilibrium is established. Then, the sample is reasonably quickly cooled to the experimental temperature. In addition to an "instantaneous" volume adjustment, one finds a substantial and often prolonged volume adjustment that takes place after the temperature jump. In Figure 4-8 this time-dependent volume change is shown for several experimental temperatures ranging from 35.0°C to 19.8°C. Here, the normalized volume departure from equilibrium, $(V - V_\infty)/V_\infty$, which we shall call δ, is plotted as a function of log time. One can clearly see that volume equilibrium is established rapidly at the higher experimental temperatures but has not been attained even after several days at 19.8°C.

The kinetic theories under discussion have as a goal the explanation for such time-dependent approaches to equilibrium. For most simple substances such as gases or single-phase liquids, the specific volume is a unique function of temperature and pressure. In the case of vitrifiable materials in the glass transition region, this is no longer so. As is clear from Figure 4-8, at 19.8°C, for example, it is possible to prepare a material with a volume almost 0.5% greater than its equilibrium volume and the sample will retain much of this volume excess for very long periods of time. Under such circumstances, rather than indicating that volume is a unique function of temperature and pressure, kinetic theories suggest writing $V(T, P, \xi)$ where ξ is called an ordering parameter and is a general measure of how far the sample is away from volume equilibrium. With the establishment of equilibrium, the volume once again becomes a unique function of temperature and pressure as ξ itself depends only on T and P at equilibrium.

At this point in the development, a differential equation is usually written. Since we are interested in the establishment of equilibrium under isothermal conditions after a single temperature jump applied to the sample initially at equilibrium, the equation is particularly simple:

$$\frac{d\delta}{dt} = -\frac{\delta}{\tau(\delta)} \qquad (4\text{-}23)$$

Once again, δ is the normalized departure from equilibrium and the equation postulates that the rate of approach toward equilibrium $(d\delta/dt)$ is proportional to δ itself. The retardation time τ may be thought of as the proportionality constant; if τ is small, equilibrium is achieved rapidly and if τ is large, slowly.

A very important aspect of kinetic theories now surfaces. Equation (4-23) is concerned with the time dependence of volume recovery after a temperature jump. For the volume to change, the molecules must move. However, as discussed in Chapter 3, the rate at which the molecules can move is a sensitive function of the free volume, and the free volume is varying with time in Figure 4-8. Thus, it is clear that the retardation time in equation (4-23) must itself depend upon δ and vary with time. This situation can be treated in quite a straightforward way by modifying equation (3-16) as

$$f = f_g + \alpha_f(T - T_g) + \delta \qquad (4\text{-}24)$$

where f, the fractional free volume, adopts equilibrium values whenever $\delta = 0$. (The usual restriction of applying equation [3–16] only to temperatures above T_g is normally relaxed at this point.) The fractional free-volume dependence of the viscosity should be the same as that of the retardation time under discussion, so that equation (3-18) may be generalized to:

$$\tau(T,\delta) = \tau(T_g,\delta=0)\exp\left(\frac{1}{f_g + \alpha_f(T - T_g) + \delta} - \frac{1}{f_g}\right) \qquad (4\text{-}25)$$

This equation explicitly relates a reference retardation time τ $(T_g,\delta=0)$ measured under conditions of volume equilibrium at T_g to a retardation time applicable under any other conditions of temperature and volume via the usual free-volume parameters.

Equations (4-23) and (4-25) have been used to calculate the solid lines that essentially connect the data points in Figure 4-8; reasonable values for the free-volume parameters and the reference retardation time were used and are listed in the figure legend.

The essentially quantitative agreement between the experimental data and the theoretical calculation which employs only three independent parameters is very satisfying. Nevertheless, other types of volume and enthalpy recovery experiments show that the single-ordering parameter model is not sufficiently flexible to rationalize many of the important aspects of the glass transition phenomenon and more complex models based on multiple-ordering parameters are now under development.

C.   STRUCTURAL PARAMETERS AFFECTING THE GLASS TRANSITION

We have seen that the free-volume, thermodynamic, and kinetic theories serve to rationalize the glass transition phenomenon in a wide variety of polymeric systems. There are, of course, additional effects that cannot be well explained on the basis of either of these approaches. In this section we examine some structural parameters affecting T_g in a qualitative manner, pointing out exceptions to the theories quoted above.

The details of the molecular structure of polymers profoundly influence the observed T_g's, as illustrated by Table 4-2, where we may contrast the T_g of polydimethyl siloxane, $-123°C$, with that of polycalcium phosphate, $+525°C$. At least approximately, we may separate the observed effects into intermolecular and intramolecular parts. The latter refer to structural parameters affecting the stiffness of the chain backbone; we shall examine these first.

The internal mobility of a chain reflects the ease of rotation of backbone bonds about one another. This is determined by the barrier to internal rotation and by steric hindrance introduced by the presence of substituents

Table 4-2. Glass Transition Temperatures for Selected Polymers

Organic Polymers	$T_g(^\circ C)$
Polyacenaphthalene	264
Polyvinyl pyrrolidone	175
Poly-*o*-vinyl benzyl alcohol	160
Poly-*p*-vinyl benzyl alcohol	140
Polymethacrylonitrile	120
Polyacrylic acid	106
Polymethyl methacrylate	105
Polyvinyl formal	105
Polystyrene	100
Polyacrylonitrile	96
Polyvinyl chloride	87
Polyvinyl alcohol	85
Polyvinyl acetal	82
Polyvinyl proprional	72
Polyethylene terephthalate	69
Polyvinyl isobutyral	56
Polycaprolactam (nylon 6)	50
Polyhexamethylene adipamide (nylon 6, 6)	50
Polyvinyl butyral	49
Polychlorotrifluorethylene	45
Ethyl cellulose	43
Polyhexamethylene sebacamide (nylon 6, 10)	40
Polyvinyl acetate	29
Polyperfluoropropylene	11
Polymethyl acrylate	9
Polyvinylidene chloride	-17
Polyvinyl fluoride	-20
Poly-1-butene	-25
Polyvinylidene fluoride	-39
Poly-1-hexene	-50
Polychloroprene	-50
Polyvinyl-*n*-butyl ether	-52
Polytetramethylene sebacate	-57
Polybutylene oxide	-60
Polypropylene oxide	-60
Poly-1-octene	-65
Polyethylene adipate	-70
Polyisobutylene	-70
Natural rubber	-72
Polyisoprene	-73

Table 4-2. (*Continued*)

Organic Polymers	$T_g(°C)$
Polybutadiene	− 85
Polydimethyl siloxane	− 123

Inorganic Polymers	$T_g(°C)$
Silicon dioxide	1200–1700
Polycalcium phosphate	525
Polysodium phosphate	285
Boron trioxide	200–260
Arsenic trisulfide	195
Arsenic trioxide	160
Zinc chloride	100
Sulfur	75
Selenium	30
Polyphosphoric acid	− 10

on the backbone chain atoms. Thus, the low T_g of polydimethyl siloxane may be rationalized on the basis of a low rotational energy barrier. A similar explanation may be advanced for the low T_g of linear polyethylene, − 128°C. (Some controversy still exists concerning the T_g of linear polyethylene. The value of − 128°C is adopted here from the recent work of Beatty and Karasz.[7]) The steric hindrance factor may be illustrated by considering the effect of the substitution of various groups for hydrogen atoms on a linear polyethylene chain. The placement of a methyl group on alternate carbon atoms results in polypropylene, which has a T_g of − 20°C. Polystyrene, which has a phenyl group on alternate carbon atoms, has a T_g of + 100°C, over 200° higher than that of unsubstituted polymethylene. This type of correlation must be applied with caution, however. Consider polyisobutylene, containing two methyl groups on alternate carbon atoms. In this case the T_g is − 80°C, some 60° lower than the monosubstituted case, polypropylene.

An increase in side group bulkiness generally serves to raise T_g. Ring-substituted polystyrenes show this effect and, in the case of polyacenaphthalene, the mobility of the chain is so severely inhibited that the T_g occurs at 264°C.

Chain microstructure is clearly important in determining chain mobility. We have already discussed the case of random copolymers on the basis of the Gibbs–DiMarzio theory and have developed equation (4-19), which shows that the T_g of a random copolymer is intermediate between the T_g's of the corresponding homopolymers. Equation (4-19) is identical in form to the two-parameter empirical expression of Wood.[6]

$$K_A W_A (T_g - T_{g_A}) + K_B W_B (T_g - T_{g_B}) = 0 \qquad (4\text{-}26)$$

In equation (4-26), W_A and W_B are the weight fractions of the comonomers A and B, and K_A and K_B are constants characteristic of A and B, respectively.

Equation (4-26) may be rearranged to

$$T_g = \frac{T_{g_A} + (kT_{g_B} - T_{g_A})W_B}{1 - (1 - k)W_B} \qquad (4\text{-}27)$$

where $k = K_B / K_A$.

Equation (4-27), known as the Gordon–Taylor[8] equation, has found wide application to random amorphous copolymers. Figure 4-9 shows the experimental results for a series of styrene–butadiene copolymers along with the corresponding T_g's calculated from equation (4-27) with $k = 0.28$.

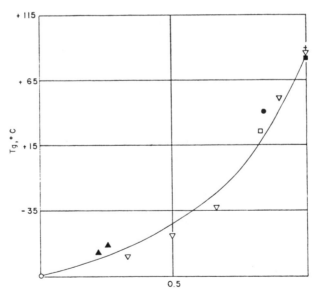

Figure 4-9. Composition dependence of T_g for a series of styrene-butadiene copolymers compared to the predicted curve calculated on the basis of equation (4-27) with $k = 0.28$. After F. Bueche et. al., *J. Polym. Sci.*, **50**, 549 (1961), by permission of John Wiley & Sons, Inc.

The above equations (4-26) and (4-27) are applicable only to random copolymers. Very different phenomena are observed in the case of block and graft copolymers. Frequently in these situations, two T_g's are observed as a consequence of the occurrence of microphase separation. This subject is beyond the scope of this book, however.

In addition to copolymer composition, geometrical and steric isomerism play important roles in the determination of chain stiffness and thus of T_g. Polydienes, for example, can exist as cis and trans geometrical isomers. In the case of poly-1,4-butadiene, the cis isomer has a T_g of $-102°C$ while the stiffer trans isomer shows a T_g of $-48°C$. The effect is not nearly so marked in the case of polyisoprene where the cis isomer exhibits a T_g of $-73°C$ and the trans isomer a T_g of $-53°C$.

Vinyl polymers of the type CH_2CHX and vinylidene polymers of the type CH_2CXY are capable of existing in different stereoregular forms. These are generally referred to as isotactic, syndiotactic, and atactic forms; it is here assumed that the reader is familiar with this nomenclature and the specific structures associated with it. It has been shown that steric isomerism has no effect on the T_g of vinyl polymers but profoundly affects the T_g of vinylidene polymers.[9] In the latter case, the isotactic form invariably is associated with the lowest T_g values. This result has also been rationalized on the basis of the Gibbs–DiMarzio theory, but again the details are beyond the scope of this book.

Turning to intermolecular effects, we first note that the free-volume approach is of great utility here, although by no means universally applicable. One example is the effect of low-molecular-weight diluents or plasticizers, which have already been accounted for by the free-volume theory through equation (4-8). The general idea is that plasticizers lower T_g by introducing free volume into the system, the final T_g being intermediate between that of the plasticizer and that of the polymer. The free-volume theory cannot account, however, for the plasticizing effect of water on polar polymers. Water exerts a dramatic plasticizing effect on nylons, greater than 30°C for the first weight per cent. However, the addition of water in low concentrations actually increases the density of the polymer–water mixture and would thus be expected to raise the T_g on the basis of the free-volume theory rather than lowering it as actually observed. The explanation is that water acts to break intermolecular hydrogen bonds present in the nylon, thus allowing greater chain mobility and also more efficient packing, the net result leading to a decrease in T_g.

The effect of molecular weight has been alluded to in Chapter 3. Specifically, free volume around chain ends is taken to be greater than that around chain middles because of imperfect packing at the chain ends. It is

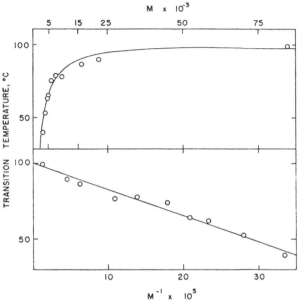

Figure 4-10. T_g versus reciprocal molecular weight for polystyrene fractions. The solid line is a plot of equation (4-28) with $c = 1.7 + 10^5$. After T. G. Fox and P. J. Flory, *J. Appl. Phys.*, **21**, 581 (1950), by permission of the American Institute of Physics.

expected that T_g will become independent of molecular weight at high molecular weights due to the negligible concentration of chain ends in high-molecular-weight polymers. It follows directly from the free-volume theory (see Problem 4-1) that

$$T_g = T_g^\infty - \frac{c}{\overline{M}_n} \tag{4-28}$$

which is known as the Fox–Flory equation.

Figure 4-10 illustrates equation (4-28) for the case of polystyrene; it is seen that the data are well fitted for $c = 1.7 \times 10^5$. The c values vary from polymer to polymer but are generally of the same order of magnitude as the polystyrene case.

D. RELAXATIONS IN THE GLASSY STATE

In addition to the glass transition, amorphous polymers usually exhibit at least one so-called secondary relaxation region. Secondary relaxations are manifestations of motions within the polymer in the glassy state. Since large-scale motions such as those accompanying the glass transition are impossible in the glassy state, the secondary relaxations must arise from localized motions. These are conveniently classified into two types: main-chain motions and side-group motions. In this section we describe methods of detecting secondary relaxations and give examples of secondary relaxations belonging to both side-group and main-chain motions.

Until recently, no secondary relaxation had been observed by calorimetry or dilatometry. It has now been reported that the side-chain relaxation in polymethyl methacrylate has been detected dilatometrically.[10] The necessary sensitivity to detect these relaxations by the dilatometric method is very difficult to achieve. These remarks also apply to the static viscoelastic measurements such as creep and stress relaxation. There is only a single report of the detection of a secondary relaxation by creep, again the side-group relaxation in polymethyl methacrylate[11] and no reports of any detected by stress relaxation.

Dynamic methods, on the other hand, easily detect glassy state relaxations and have been extensively applied to their study. These include dynamic mechanical methods, dielectric relaxation, and nuclear magnetic resonance. Since we are primarily concerned with viscoelastic response at this point, we shall confine the discussion to the dynamic mechanical technique and delay our consideration of dielectric relaxation until Chapter 8. As should be apparent, it is possible to detect secondary relaxations by performing measurements over a frequency range at constant temperature or over a temperature range at constant frequency. The latter technique is usually employed, largely because extended mechanical frequency ranges are difficult to achieve experimentally and many secondary relaxations of interest occur at very high frequencies at room temperature.

As an example may be cited the case of polymethyl methacrylate. Figure 4-11 shows a plot of the logarithmic decrement versus temperature for this polymer at a constant frequency of 1 cps in the temperature range $-50°$ to $+160°C$. Two relaxation peaks are discernible, the higher temperature relaxation corresponding to the glass transition ($T_g \approx 105°C$) and a secondary relaxation at 50°C. It becomes necessary to adopt a labeling scheme for the various relaxations observed. Several schemes are currently in use and we shall choose one of the most popular, that of labeling the peaks α, β, γ, and so on, in order of decreasing temperature. Thus, in a linear amorphous polymer, the α relaxation corresponds to the glass transition, the β relaxation to the highest temperature glassy-state relaxation, and so forth. Polymethyl methacrylate exhibits only one secondary relaxation in the temperature range of Figure 4-11 and this is the β relaxation at 50°C.

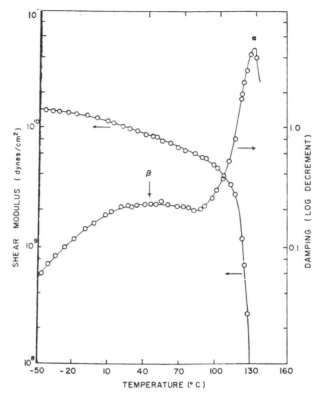

Figure 4-11. Temperature dependence of tan δ for polymethyl methacrylate showing the α and β relaxations. After L. E. Nielsen, *Mechanical Properties of Polymers*, Reinhold, New York, 1962, p. 178.

Much of the discussion of the motions underlying the glass transition has been presented without reference to details of molecular structure. Because of the relatively long-range motions characterizing the glass transition, it is possible to construct general theories applicable to all amorphous polymers without regard to fine structural details. Such is not the case with the localized secondary relaxations. We must in general consider structural features concerning only a few main chain or side group atoms. Many different mechanisms have been proposed for the large variety of secondary relaxations observed, and a detailed consideration is beyond the scope of this treatment. A very complete account exists in the work of McCrum, Read, and Williams,[12] which should be consulted for more details. We shall present only two mechanisms here.

Figure 4-12.   The crankshaft mechanism according to Schatzki.

First is the crankshaft mechanism of Schatzki.[13] It has been observed for many polymers containing linear $(CH_2)_n$ sequences with $n = 4$ or greater, that a secondary relaxation occurs at about $-120°C$ at 1 cps. This seems to be true regardless of whether the CH_2 sequences occur in the main chain or in the side groups. Thus both polyethylene and poly-n-butyl methacrylate exhibit this relaxation. The mechanism proposed by Schatzki is shown in Figure 4-12.

The motion responsible for the relaxation is a rotation about the two colinear bonds 1 and 7 such that the carbon atoms between bonds 1 and 7 move in the manner of a crankshaft. The colinearity of the two terminal bonds is achievable if there are four intervening carbon atoms on the assumption of tetrahedral valence angles and a rotational isomeric state model. Support is to be found for the crankshaft mechanism in the fact that the activation energy estimated for the model, 13 kcal mole^{-1}, is close to the experimental results, 12–15 kcal mole^{-1}, and in the fact that the predicted free volume of activation, about four times the molar volume of a CH_2 unit, is also in good agreement with experimental estimates based on pressure studies.

An outstanding example of the identification of a secondary relaxation with a specific molecular structure lies in the work of Heijboer.[14] Heijboer studied a large number of methacrylate polymers containing the cyclohexyl group in the ester side chain. His results for a series of methyl methacrylate-cyclohexyl methacrylate copolymers are shown in Figure 4-13. It can be seen that the magnitude but not the temperature of the low temperature relaxation is affected by the concentration of cyclohexyl groups. The observed activation energy for this relaxation is 11.5 kcal/mole, identical with that observed for a mechanical relaxation in many low-molecular-weight cyclohexyl derivatives. This led to the identification of the polymeric relaxation with a flipping of the substituted cyclohexyl ring from one chain conformation to another.

The time–temperature or frequency–temperature superposition scheme discussed in Chapter 3, Section B, is applicable to secondary relaxations as well as to the glass transition, assuming that the observed secondary relaxation peaks are well resolved. When the shift factors are obtained for

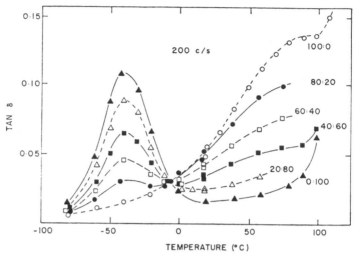

Figure 4-13. Temperature dependence of tan δ for copolymers of methyl methacrylate and cyclohexyl methacrylate. Open circles are pure polymethyl methacrylate. After J. Heijboer, *Kolloid-Z.*, **171**, 7 (1960), by permission of Dr. Dietrich Steinkopff Verlag.

these secondary relaxations, it is found that they do not obey the WLF equation but an equation of the Arrhenius form, that is:

$$\log a_T \propto -\frac{H_a}{2.303\,RT} \tag{4-29}$$

Thus plots of $\log a_T$ versus $1/T$ for a secondary relaxation will yield straight lines, not curves as in the WLF case. This fact has been used to distinguish the main glass transition from other relaxations occurring in semicrystalline polymers.

PROBLEMS

1. Assuming that the molecular-weight dependence of the fractional free volume is given by

$$f_M = \frac{f_\infty + A}{\overline{M}_n}$$

where f_M is the fractional free volume at molecular weight M, f_∞ is the fractional free volume at infinite molecular weight, and A is a constant, derive equation (4-28).

2. It has been observed that methacrylate polymers containing linear $(CH_2)_n$ sequences in the ester side chain exhibit a regular decrease in T_g with increasing n. Devise an explanation for this on the basis of the free-volume theory.

3. Qualitatively predict the dependence of T_g on strain on the basis of

    (a) The free-volume theory

    (b) The Gibbs–DiMarzio theory

Might your results suggest an experimental test to distinguish between the two theories? What difficulties might be expected in applying such a test?

4. (a) If we consider a polymer to be made of only chain ends and chain middles, show that the following equation may be obtained from equation (4-28):

$$T_g = T_g^\infty - \frac{cw_e}{M_e}$$

where w_e is the weight fraction of the chain ends and M_e is the weight of chain ends per mole of chains.

    (b) One of the equations used to predict glass transition temperature depression due to the incorporation of solvent is

$$T_g = T_{g_p} w_p + T_{g_s} w_s + K w_p w_s$$

where subscripts p and s refer to the polymer and the solvent, respectively, and K is a constant. Rearrange this equation into a form similar to the one in part (a) by considering chain ends as solvent.

5. (a) For a compatible blend (mixture) of polymer 1 with polymer 2, derive the following formula for T_g, the glass transition temperature of the blend:

$$\ln T_g = \frac{m_1(\Delta C_{P1})\ln T_1 + m_2(\Delta C_{P2})\ln T_2}{m_1(\Delta C_{P1}) + m_2(\Delta C_{P2})}$$

m_1 and m_2 are mass fractions, T_1 and T_2 are glass transition temperatures and ΔC_{P1}, and ΔC_{P2} are heat capacity differences (per unit mass) between the liquid and glassy states for polymers 1 and 2 respectively.

You may assume that the entropy of the blend is given as

$$S = X_1 S_1 + X_2 S_2$$

where X denotes mole fraction and that at all glass transition temperatures,

$$S_{\text{glass}} = S_{\text{liquid}}$$

    (b) Also show that if T_1 and T_2 are close and $\Delta C_{P1} = \Delta C_{P2}$, the above relationship reduces to:

$$T_g = m_1 T_1 + m_2 T_2$$

[See P. B. Couchman, *Phys. Lett.*, **70A**, 155 (1979).]

6. In order to roughly assess the importance of the variation of retardation time with recovery in the single-ordering-parameter kinetic view of the glass transition, contrast the experimentally observed behavior given in Figure 4-8 with that predicted by a single-ordering parameter model with a structurally independent retardation time:

$$\frac{d\delta}{dt} = -\frac{\delta}{\tau}$$

Values of $\delta_0 = 3.0 \times 10^{-3}$ and $\tau = 1.0$ hours are appropriate for consideration of the 24.9°C curve in the figure.

REFERENCES

1. R. S. Porter and J. F. Johnson, Eds., *Analytical Calorimetry*, Plenum, New York, 1969.
2. P. Ehrenfest, *Leiden Comm. Suppl.*, 756 (1933).
3. W. Kauzmann, *Chem. Rev.*, **43**, 219 (1948).
4. J. H. Gibbs and E. A. DiMarzio, *J. Chem. Phys.*, **28**, 373 (1958).
5. R. Simha and R. F. Boyer, *J. Chem. Phys.*, **37**, 1003 (1962).
6. L. A. Wood, *J. Polym. Sci.*, **28**, 319 (1958).
7. C. L. Beatty and F. E. Karasz, *J. Macromol. Sci., Rev. Macromol. Chem.*, **C17**, 37 (1979).
8. M. Gordon and J. S. Taylor, *J. Appl. Chem.*, **2**, 493 (1952).
9. F. E. Karasz and W. J. MacKnight, *Macromolecules*, **1**, 537 (1968).
10. R. A. Haldon, W. J. Schell, and R. Simha, *J. Macromol. Sci.—Phys.*, **B1**, 759 (1967).
11. W. Lethersich, *Brit. J. Appl. Phys.*, **1**, 294 (1950).
12. N. G. McCrum, B. E. Read, and G. Williams, *Anelastic and Dielectric Effects in Polymeric Solids*, Wiley, New York, 1967.
13. T. F. Schatzki, *J. Polym. Sci.*, **57**, 496 (1962).
14. J. Heijboer, *Kolloid-Z.*, **171**, 7 (1960).

5

Statistics of a Polymer Chain

The special structure of polymer molecules that distinguishes them from other species is their long, flexible chain nature, which has already been discussed in Chapter 1. Such characteristic properties as long-range elasticity and high viscosity are direct consequences of this special structure. A qualitative overview of the viscoelastic response of polymers and the molecular interpretation of this response has been presented in Chapter 3. In order to provide a quantitative description of the viscoelastic behavior of polymers on the basis of their molecular structure (as opposed to the phenomenological description presented in Chapter 2), let us first consider an isolated polymer chain and then extend the results to ensembles of chains, that is, to the bulk polymer itself. An isolated linear polymer chain is capable of assuming many different conformations. Because of thermal agitation, such a chain is also constantly changing from one conformation to another, namely, it is undergoing Brownian motion. It is, of course, impractical, if not impossible, to analyze all possible conformations in detail. We may only hope to obtain various average properties for the chain by the application of statistical methods.

Because of the presence in real chains of short- and long-range interactions between segments, bond angle restrictions, potential energy barriers, and other characteristics, an exact statistical calculation becomes a very complex mathematical problem. However, because these chains have such a large number of conformations, idealized models can be used to derive the average properties. In most cases, these quantities provide a good asymptotic approximation to the true values for any sufficiently long linear polymer chain.

We shall first derive the average properties of an ideal polymer chain that is infinitely long, possesses negligibly small volume, and has freely jointed links. Next we shall examine the influence of fixed bond angles between adjacent links. The concept of the statistically equivalent random chain will then be introduced to rationalize the validity of using these model chains to represent the behavior of real polymer chains. Finally, the equation of state for a single polymer chain will be discussed.

A. THE MEAN SQUARE END-TO-END DISTANCE

Suppose our ideal polymer chain has n links, each of length l, then the fully extended length of the chain would be:

$$R = nl \tag{5-1}$$

However, since, as we have just mentioned, the fully extended conformation is only one of a great many, it would be more meaningful to consider an average size of the macromolecule such as the mean square end-to-end distance, $\overline{r^2}$. As the name implies, the end-to-end distance is just the length of the vector connecting the two ends of the ideal chain. The average we seek can be considered to be either the average of the values for a given molecule at a number of times or that of an ensemble of identical molecules at the same time.[†] Thus, for an ensemble of chains that do not interact with one another,

$$\overline{r^2} = \frac{1}{p} \sum_{i=1}^{p} \mathbf{r}_i^2 \tag{5-2}$$

where $\mathbf{r}_i$ refers to the end-to-end distance of the ith chain, and p to the number of chains. Now let $\mathbf{l}_j$ designate the vectorial length of link j, then

$$\mathbf{r}_i = \sum_{j=1}^{n} \mathbf{l}_j = \mathbf{l}_1 + \mathbf{l}_2 + \mathbf{l}_3 + \cdots + \mathbf{l}_n \tag{5-3}$$

[†]For a justification of this assertion, the reader is referred to standard statistics texts such as R. C. Tolman, *Principles of Statistical Mechanics*, Oxford, 1938.

Equation (5-2) now becomes

$$\overline{r^2} = \frac{1}{P} \sum_{i=1}^{P} \left(\sum_{j=1}^{n} \mathbf{l}_j \right)_i^2 \tag{5-4}$$

The square of the end-to-end distance of a particular chain is obtained from equation (5-3).

$$r_i^2 = \mathbf{r}_i \cdot \mathbf{r}_i$$

$$= \sum_{j=1}^{n} \mathbf{l}_j \cdot \mathbf{l}_j + 2 \sum_{k<j} \mathbf{l}_k \cdot \mathbf{l}_j \tag{5-5}$$

But from vector algebra we know that the dot product of two vectors is just the product of their absolute values times the cosine of the angle, θ, between one vector and the other vector. Since all the links in our idealized chain are identical in length we have

$$\mathbf{l}_k \cdot \mathbf{l}_j = |\mathbf{l}_k||\mathbf{l}_j|\cos \theta_{kj}$$

$$= l^2 \cos \theta_{kj} \tag{5-6}$$

The angle between the same vectors $\mathbf{l}_j$ in the first sum of equation (5-5) is zero, and the cosine of zero is unity. Thus the first sum in equation (5-5) is simply

$$\sum_{j=1}^{n} \mathbf{l}_j \cdot \mathbf{l}_j = nl^2 \tag{5-7}$$

Now in the second sum of equation (5-5) we deal with the product of two different vectors $\mathbf{l}_k$ and $\mathbf{l}_j$. We recall that our model chain is freely jointed and volumeless. Thus any two links can assume any orientation whatsoever with respect to each other, including interpenetrating into one another. In other words, any angle θ between two vectors $\mathbf{l}_k$ and $\mathbf{l}_j$ is just as probable as any other angle. Thus for every $\cos \theta$ in the second term of equation (5-5) there is a $\cos(\pi + \theta) = -\cos \theta$ that exactly cancels it. The result is, of course, that the second term in equation (5-5) is 0. This gives us the mean square end-to-end distance of a freely orienting chain[1-4]:

$$\overline{r^2} = \frac{1}{P} \sum_{i=1}^{P} nl^2$$

$$= nl^2 \tag{5-8}$$

In addition, it is possible to calculate the property $\overline{x^2}$, i.e., the average of the square of the projection of $\mathbf{r}_i$ on to any fixed axis, in this case the x axis. Here we may proceed by considering the individual conformations accessible to the chain as before, or we may realize that x^2 is given, in general, as an average over a distribution function

$$\overline{x^2} = \int_{-l}^{l} x^2 p(x) \, dx \tag{5-9}$$

$p(x) \, dx$ is the probability that x is between x and $x + dx$. To calculate $\overline{x^2}$ we must first determine $p(x)$. The vector $\mathbf{r}_i$ on the average has a length $n^{1/2}l$, and since the entire chain is freely orienting, $\mathbf{r}_i$ itself is evenly distributed through all possible directions. Thus the locus of possible

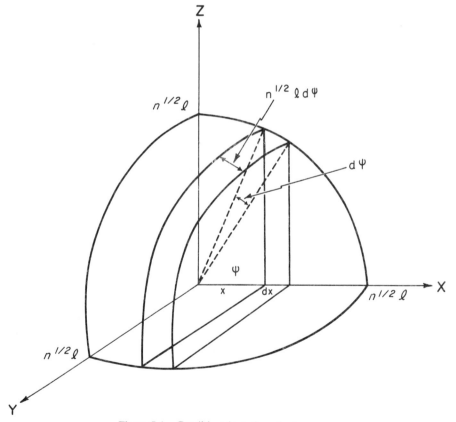

Figure 5-1.   Possible orientation of **r** in space.

average configurations of $\mathbf{r}_i$ is given as the sphere of radius $n^{1/2}l$, one octant of which is shown in Figure 5-1. Now to observe a certain value of x, say between x and $x + dx$, it should be clear that $\mathbf{r}_i$ would have to fall in the spherical segment shown. The probability that $\mathbf{r}_i$ will fall in this segment is just equal to the area of this segment, $2\pi(n^{1/2}l)^2\sin\psi\,d\psi$ divided by the total surface area of the sphere $4\pi(n^{1/2}l)^2$. Thus,

$$p(x)\,dx = \tfrac{1}{2}\sin\psi\,d\psi \qquad (5\text{-}10)$$

and using equation (5-9)

$$\overline{x^2} = \tfrac{1}{2}\int_0^\pi (n^{1/2}l)^2\cos^2\psi\,\sin\psi\,d\psi$$

$$= \frac{nl^2}{3} = \frac{\overline{r^2}}{3}$$

B.   STATISTICAL DISTRIBUTION OF END-TO-END DISTANCES

Equation (5-8) tells us that the root mean square length of a freely orienting chain is proportional to the square root of the number of links in the chain. Now we ask: Given a chain of n links, each of which has a length l, what is the probability of achieving a particular end-to-end length? To be more precise, if one end of this chain is at the origin of a Cartesian coordinate system, what are the chances of finding the other end in a small volume element $dx\,dy\,dz$ which is at a distance $\mathbf{r}$ from the origin (Figure 5-2)? This problem is known as the random flight problem and will be solved for the one-dimensional case first.[5-8] We thus constrain $\mathbf{r}$ to lie on the x-axis, then calculate the probability $\omega(x)$ that x has a value between x and $x + dx$. For a chain consisting of a large number of links, we can assume that every link contributes a length $l/\sqrt{3}$ [equation (5-10)] to the component on the x-axis. Some of them will be in the positive x direction (n_+ of them in number) and some in the negative (n_-). Since the chain is freely jointed, the probability of taking either direction will be equal, namely $1/2$. Obviously, the total length of the x-axis component is the number of components in the positive direction minus that in the negative direction multiplied by the average length of each component:

$$x = \frac{(n_+ - n_-)l}{\sqrt{3}} \qquad (5\text{-}11)$$

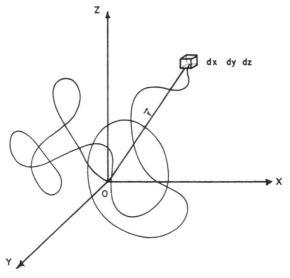

Figure 5-2. Conformation of a polymer chain with one end fixed at the origin of a Cartesian coordinate system.

Now the probability of a random flight with n_+ positive steps and n_- negative steps is:[9]

$$\omega(n_+, n_-) = \frac{(1/2)^n n!}{n_+! \, n_-!} \tag{5-12}$$

where

$$n = n_+ + n_- \tag{5-13}$$

Let

$$m = n_+ - n_- \tag{5-14}$$

Equation (5-12) becomes

$$\omega(m, n) = \frac{(1/2)^n n!}{[(n+m)/2]! \, [(n-m)/2]!} \tag{5-15}$$

If $n \gg m$, we can use Stirling's approximation[†]

$$\ln n! = n \ln n - n \tag{5-16}$$

for $n \to \infty$.

Therefore, after considerable arithmetic rearrangement,

$$\ln[\omega(m,n)] = -\frac{n+m}{2}\ln\left(1+\frac{m}{n}\right) - \frac{n-m}{2}\ln\left(1-\frac{m}{n}\right) \tag{5-17}$$

However, since $n \gg m$, we can use the following series expansion:

$$\ln\left(1 \pm \frac{m}{n}\right) = \pm\frac{m}{n} - \frac{m^2}{2n^2} \tag{5-18}$$

In equation (5-18) third-order and higher-order terms of m/n are neglected. We thus have

$$\ln[\omega(m,n)] = -\frac{m^2}{2n} \tag{5-19}$$

In using Stirling's approximation, some small terms which comprise the normalization of $\omega(m,n)$ have been neglected so that this probability must be renormalized before we can proceed. To do this, we may write

$$1 \equiv K \sum_{m=-n}^{m=n} \omega(n,m) \tag{5-20a}$$

where the probability of m having any value between $-n$ and $+n$ must be unity. We note that the minimum change in Δm is 2, namely, for one bond to change from a positive to a negative orientation or vice versa. Thus, in reality, equation (5-20a) should be normalized for values of $m = -n$, $-n+2$, $-n+4, \ldots, n-2$, n. Mathematically it is simpler to normalize this function over all integer values of m between $-n$ and $+n$ and to realize that this normalization counts twice as many terms and thus should be normalized to 2.0 instead of to 1.0. Thus we rewrite equation (5-20a) as

$$2 \equiv K \sum_{m=-n}^{m=n} \omega(n,m) \tag{5-20b}$$

[†] $n! = 1 \times 2 \times 3 \cdots (n-1) \times n$

$\ln n! = \ln 1 + \ln 2 \cdots \ln(n-1) + \ln n = \sum_{x=1}^{n} \ln x$

$\qquad \approx \int_1^n \ln x \, dx = [x \ln x - x]_1^n \approx n \ln n - n$

where m takes on all integer values. If n is large, equation (5-20b) may be approximated by the integral

$$K \int_{-n}^{n} \omega(n, m)\, dm = 2 \tag{5-21}$$

Substitution of the nonnormalized equation (5-19) into (5-21) yields upon variable transformation, $(y = m/\sqrt{2n})$

$$\sqrt{2n}\, K \int_{-n}^{n} e^{-y^2} dy = 2 \tag{5-22}$$

If n is sufficiently large,

$$\sqrt{2n}\, K \int_{-\infty}^{\infty} e^{-y^2} dy = 2 \tag{5-23}$$

which is merely

$$\sqrt{2n}\, K \left(2 \int_{0}^{\infty} e^{-y^2} dy \right) = 2 \tag{5-24}$$

The term in brackets in this equation is just the gamma function of argument $1/2$, which is equal to $\sqrt{\pi}$. Thus:

$$K = \sqrt{\frac{2}{n\pi}} \tag{5-25}$$

and the normalized probability $\omega(n, m)$ is given as

$$\omega(n, m) = \sqrt{\frac{2}{n\pi}}\, e^{-m^2/2n} \tag{5-26}$$

However, $\omega(m, n)$ is equal to the probability $\omega(x)\Delta x$ that the length lies between x and $x + \Delta x$:

$$\omega(m, n) = \omega(x)\Delta x \tag{5-27}$$

Once again, note that the minimum change in m is 2. Thus Δx corresponds to $\Delta m = 2$. By substituting $\Delta x = 2l/\sqrt{3}$, and $m = \sqrt{3}x/l$ [from equation (5-11)], we get:

$$\omega(x)\, dx = \left(\frac{3}{2\pi nl^2} \right)^{1/2} \exp\left(\frac{-3x^2}{2nl^2} \right) dx \tag{5-28}$$

Equation (5-28) is known as the Gaussian distribution function. It can be generalized to the three-dimensional case as follows:

$$w(x, y, z) \, dx \, dy \, dz = w(x) \, dx \, w(y) \, dy \, w(z) \, dz$$

$$= \left(\frac{b}{\pi^{1/2}} \right)^3 \exp(-b^2 r^2) \, dx \, dy \, dz \qquad (5\text{-}29)$$

where

$$b^2 = \frac{3}{2nl^2} = \frac{3}{2\overline{r^2}}$$

$$r^2 = x^2 + y^2 + z^2$$

Equation (5-29) gives the probability that if one end of a freely orienting chain is fixed at the origin, the other end will be found in the volume element $dx \, dy \, dz$ located at $\mathbf{r}$ away from the origin.

Rather than the probability of finding the free chain end in a particular volume element $dx \, dy \, dz$, we are often more interested in knowing the probability of finding the free chain end in a spherical shell of radius r centered at the origin. (Here r is not a vector quantity.) In this case, the appropriate function is the so-called radial distribution function $w(r) \, dr$ obtained by multiplying $w(x, y, z)$ by the volume of a spherical shell of thickness dr located at a distance r from the origin. This volume is $4\pi r^2 dr$:

$$w(r) \, dr = \left(\frac{b}{\pi^{1/2}} \right)^3 \exp(-b^2 r^2) 4\pi r^2 dr \qquad (5\text{-}30)$$

Equation (5-30) is illustrated in Figure 5-3.

We note that in the distribution curve there is a maximum that corresponds to the most probable end-to-end distance. We can find the position of this maximum by differentiating equation (5-30) with respect to r:

$$\frac{dw(r)}{dr} = 8\pi r \left(\frac{b}{\pi^{1/2}} \right)^3 (1 - b^2 r^2) \exp(-b^2 r^2) \qquad (5\text{-}31)$$

Setting equation (5-31) equal to zero and solving, the most probable value of r is $1/b$ or $(2nl^2/3)^{1/2}$. The mean square end-to-end distance is the second moment of the radial distribution function and is given by:

$$\overline{r^2} = \frac{\int_0^\infty r^2 w(r) \, dr}{\int_0^\infty w(r) \, dr} \qquad (5\text{-}32)$$

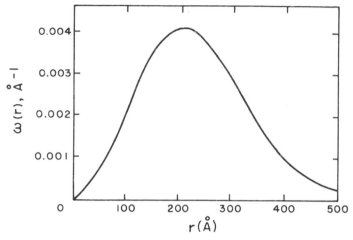

Figure 5-3. Radial distribution function of a chain consisting of 10^4 freely orienting segments each of length 2.5 Å. After P. J. Flory, *Principles of Polymer Chemistry*, Cornell University Press, Ithaca, N.Y., 1950.

However, the denominator of the right side of equation (5-32) is unity, since the radial distribution function is normalized. Integration of equation (5-32) yields:

$$\overline{r^2} = \tfrac{3}{2}b^{-2} = nl^2 \qquad (5\text{-}33)$$

which is just the expression we have derived in the previous section [equation (5-8)].

C. EFFECT OF BOND ANGLE RESTRICTIONS

In the preceding two sections, we have treated the conformational statistics of an idealized model chain. In real polymers the situation is, of course, considerably more complicated. However, for the sake of mathematical tractability, the quantities derived for the model chain will be used for the development of molecular theories of elasticity and viscoelasticity of polymers to be presented in the later chapters. But it is instructive to consider the effects of incorporating some of the features of a real polymer chain into the model. As an example, we shall treat in this section the effect of bond angle restrictions on the statistics of a polymer chain.

We consider a freely rotating polymer chain whose jth segment orientation is determined by the bond angle it forms with its immediate neighbor, $(j-1)$th segment (Figure 5-4). Combining equations (5-4), (5-5), and (5-7),

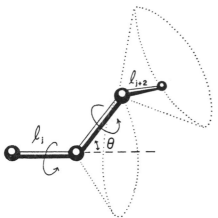

Figure 5-4. Segment of a freely rotating chain with fixed bond angles.

we have

$$\overline{r^2} = nl^2 + \frac{1}{p}\sum_{i=1}^{p}\left(\sum_{k\neq j}^{n} \mathbf{l}_k \cdot \mathbf{l}_j\right)_i \qquad (5\text{-}34)$$

In this case, however, angles between successive segments cannot assume any value as in the case of the freely orienting chain (no bond angle restrictions) in the preceding section. They must have a fixed value. The crossterm in equation (5-34) thus will not vanish. To evaluate this term, we first consider the dot product $\mathbf{l}_j \cdot \mathbf{l}_{j+1}$ in the sum. The projection of $\mathbf{l}_j$ onto its nearest neighbor is $l\cos\theta$. Thus $\mathbf{l}_j \cdot \mathbf{l}_{j+1} = l^2\cos\theta$; it is obvious that this value is also equal to the ensemble average of $\mathbf{l}_j \cdot \mathbf{l}_{j+1}$. Now focus on next nearest neighbors and the term $\mathbf{l}_j \cdot \mathbf{l}_{j+2}$. With respect to Figure 5-4, one can see that the value of this dot product will fluctuate as $\mathbf{l}_{j+2}$ freely rotates about its axis described by $\mathbf{l}_{j+1}$. Nevertheless, the average projection of $\mathbf{l}_{j+2}$ onto $\mathbf{l}_{j+1}$ is $l\cos\theta$ and, since the average projection of a vector of magnitude $l\cos\theta$ pointing in the same direction as $\mathbf{l}_{j+1}$ onto the vector $\mathbf{l}_j$ will be $l\cos^2\theta$, the ensemble average of $\mathbf{l}_j \cdot \mathbf{l}_{j+2}$ is given as $l^2\cos^2\theta$. Accordingly for other terms in the summation, equation 5-34 becomes:

$$\overline{r^2} = nl^2 + (\cos\theta + \cos^2\theta + \cos^3\theta + \cdots + \cos^{n-1}\theta)l^2$$

$$+ (\cos\theta + \cos\theta + \cos^2\theta + \cdots + \cos^{n-2}\theta)l^2$$

$$+ (\cos^2\theta + \cos\theta + \cos\theta + \cdots + \cos^{n-3}\theta)l^2$$

$$+ \cdots\cdots\cdots$$

$$+ (\cos^{n-1}\theta + \cos^{n-2}\theta + \cos^{n-3}\theta + \cdots + \cos\theta)l^2 \qquad (5\text{-}35)$$

Collecting $\cos\theta$ terms,

$$\overline{r^2} = nl^2 + \left[2(n-1)\cos\theta + 2(n-2)\cos^3\theta \right.$$

$$\left. + 2(n-3)\cos^3\theta + \cdots + 2\cos^{n-1}\theta \right]l^2$$

$$= nl^2 + 2nl^2\cos\theta(1 + \cos\theta + \cos^2\theta + \cdots + \cos^{n-2}\theta)$$

$$- 2l^2\cos\theta(1 + 2\cos\theta + 3\cos^2\theta + \cdots + n\cos^{n-3}\theta) \qquad (5\text{-}36)$$

The series in equation (5-36) can be summed to yield

$$\overline{r^2} = nl^2 + 2nl^2 \frac{\cos\theta(1 - \cos^n\theta)}{1 - \cos\theta}$$

$$- 2l^2 \frac{\cos\theta(1 - \cos^n\theta)}{(1 - \cos\theta)^2} \qquad (5\text{-}37)$$

The former sum is just a geometric series while the latter series may be generated by the MacLaurin expansion of $(1 - x^n)/(1 - x)^2$. Here we have let $n - 2 \approx n - 1 \approx n$. Since n is so large, however, the third term on the right side of equation (5-37) can be neglected. $\cos\theta$ is less than 1, thus $\cos^n\theta$ can also be neglected. Equation (5-37) reduces to

$$\overline{r^2} \approx nl^2 \left[1 + \frac{2\cos\theta}{1 - \cos\theta} \right] \qquad (5\text{-}38)$$

or

$$\overline{r^2} \approx nl^2 \frac{1 + \cos\theta}{1 - \cos\theta} \qquad (5\text{-}39)$$

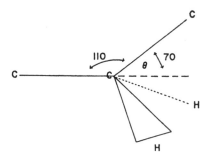

Figure 5-5. Segment of a polymethylene chain showing the tetrahedral bond angles.

For a polymethylene chain, which has tetrahedral bond angles, the angle between a given segment and its projection on the next segment is about 70° (Figure 5-5). Thus, $\cos\theta = 0.34$, and

$$\overline{r^2} = 2nl^2 \tag{5-40}$$

which is twice the value of a freely orienting chain.

D.   THE STATISTICALLY EQUIVALENT RANDOM CHAIN

We have seen in the preceding sections that drastic approximations have been introduced in our treatment of the statistics of a polymer chain. We must therefore provide ourselves with some justification for representing real chains with these simplified models. A useful concept for this purpose is the statistically equivalent random chain.[10] For a sufficiently long real chain, we can find a random chain with the same mean square end-to-end distance as the real chain:

$$\overline{r^2} = \overline{r_e^2} = n_e l_e^2 \tag{5-41}$$

In equation (5-41) subscript e refers to the equivalent random chain. However, if both the number and the lengths of segments of the equivalent random chain can be chosen at will, there would be an infinite selection. Thus we must impose another condition; this is chosen so that the fully extended length of the equivalent random chain must be equal to that of the real chain without the distortion of bond angles and lengths:

$$R = R_e = n_e l_e \tag{5-42}$$

Thus there is only one model chain that is statistically equivalent to the real chain with

$$n_e = \frac{R^2}{\overline{r^2}} ; \qquad l_e = \frac{\overline{r^2}}{R} \tag{5-43}$$

By implication, the real chain then obeys the same Gaussian distribution function as the model chain.

As an example, we consider a polymethylene chain (Figure 5-6). Its fully

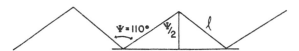

Figure 5-6. Polymethylene backbone in the fully extended conformation.

extended length and mean square end-to-end distance are respectively (considering only bond angle restrictions)

$$R = nl \sin \frac{\psi}{2} = 0.82nl \qquad (5\text{-}44)$$

$$\bar{r}^2 = 2nl^2 \qquad (5\text{-}45)$$

Combining with equations (5-43) we get

$$n_e = 0.33n \qquad l_e = 2.4l \qquad (5\text{-}46)$$

Thus each statistically equivalent random segment is equal to three carbon–carbon bonds in the real polymethylene chain.

The effect of tetrahedral bond angles on the radial distribution function enters through the parameter b, the value of which decreased from $3/2nl^2$ in the freely orienting chain to $3/4nl^2$ in the chain with tetrahedral bond angles.

E. THE EQUATION OF STATE FOR A POLYMER CHAIN

We have thus far discussed the statistical properties of a single chain that has no external constraints imposed on it. If one end of this polymer chain is fixed at the origin of a coordinate system, then due to Brownian motion the position of the other end of the chain will fluctuate according to the Gaussian distribution function [equation (5-30)]. In general, there are a large number of conformations that the chain can assume consistent with its ends being separated by a distance r. The number of conformations for each r is proportional to the radial distribution function. If we constrain the chain ends to remain a fixed distance r apart, conformations consistent with all other end-to-end distances become unavailable. As a consequence, the "degree of randomness" is now lessened—in other words, the entropy is decreased. A tension must therefore be set up owing to this perturbation. Let us now see how this tension can be related to the chain dimensions and to the magnitude of the Brownian motion, which is determined by the thermal energy available to the chain (temperature).

We recall from the first law of thermodynamics that

$$dU = T\,dS - dW \qquad (5\text{-}47)$$

where U is internal energy, T is temperature, S is entropy, and W is work done by the system. By definition, the Helmholtz free energy is

$$dA = dU - T\,dS \qquad (5\text{-}48)$$

at constant temperature. Equations (5-47) and (5-48) thus tell us that

$$dA = -dW \qquad (5\text{-}49)$$

for an isothermal process where the stress-strain work is just

$$dW = -f\,d\mathbf{r} \qquad (5\text{-}50)$$

Thus the tensile force on a polymer chain at constant temperature and length is just

$$f = -\left(\frac{\partial W}{\partial \mathbf{r}}\right)_T = +\left(\frac{\partial A}{\partial \mathbf{r}}\right)_T$$
$$= \left(\frac{\partial U}{\partial \mathbf{r}}\right)_T - T\left(\frac{\partial S}{\partial \mathbf{r}}\right)_T \qquad (5\text{-}51)$$

In the case of our model chain, which has no energy barrier hindering the rotation of the segments, the internal energy of the chain is the same for all conformations. Thus the first term on the right side of equation (5-51) is zero, and

$$f = -T\left(\frac{\partial S}{\partial \mathbf{r}}\right)_T \qquad (5\text{-}52)$$

To calculate the entropy for our Gaussian chain, we use the Boltzmann's relation from statistical mechanics:

$$S = k \ln \Omega \qquad (5\text{-}53)$$

where k is Boltzmann's constant and Ω is the total number of conformations available to the system. In our case, we want Ω to be a function of $\mathbf{r}$ the vector separation of the chain ends. Let there be $\mathfrak{N}$, some large fixed number of conformations available to a chain. Then the number of conformations consistent with a certain $\mathbf{r}$ is just

$$\Omega(\mathbf{r}) = \mathfrak{N}\omega(\mathbf{r}) \qquad (5\text{-}54)$$

where $\omega(\mathbf{r})$ is just $\omega(xyz)$ of equation (5-29). Thus,

$$\Omega(\mathbf{r}) = \mathfrak{N}\left(\frac{b}{\pi^{1/2}}\right)^3 e^{-b^2 r^2} \qquad (5\text{-}55)$$

Inserting equation (5-55) into equation (5-53) and differentiating according to equation (5-52), we obtain the equation of state for a single polymer chain:[9,11–14]

$$f = 2kTb^2\mathbf{r} \tag{5-56}$$

In equation (5-56), f is also a vector. Thus if the ends of a chain are held a fixed distance r apart by a tensile force, then this force is directly proportional to r, and to absolute temperature. It is also inversely proportional to the mean square length of the chain, since $b^2 = 3/2\overline{r^2}$ [equation (5-29)]. In fact, equation (5-56) is just Hooke's law for a spring with modulus $2kTb^2$. The elasticity of this spring originates in the decrease in conformational entropy with chain end separation and thus it is often referred to as an "entropy spring." We show in Chapter 7 that this entropy spring concept is basic to the formulation of the molecular theory of polymer viscoelasticity. If, instead of a single chain, we have a network of chains, an analogous derivation will give us the equation of state for rubber elasticity, which is the subject of the next chapter.

PROBLEMS

1. Write out in detail equation (5-2) with the aid of equations (5-3)–(5-5) if $n = 4$, $p = 1$.

2. The radial distribution function [equation (5-29)] was derived for a chain possessing negligible volume. Suppose we now have a chain whose segments have excluded volume (no two segments can occupy the same space), how would the most probable distance (the maximum in Figure 5-3) be affected?

3. Carry out the integrations indicated in equation (5-32). *Hint*:

$$\int_0^\infty x^{2a} e^{-bx^2} dx = \frac{1 \cdot 3 \cdot 5 \cdot \cdots \cdot (2a-1)}{2^{a+1} b^a} \sqrt{\frac{\pi}{b}}$$

4. Write out in detail equation (5-35) if $n = 4$.

5. It is known that for a poly-*cis*-isoprene chain, each monomer unit is 4.6 Å and $\overline{r^2} = 16.2n$. What is the statistically equivalent random chain for this macromolecule?

6. Using the following more exact statement of Stirling's approximation,

$$\ln n! = (n + \tfrac{1}{2})\ln(n) - n + \tfrac{1}{2}\ln 2\pi$$

derive equation (5-26) from equation (5-15) and show that in this case renormalization is not necessary.

7. Calculate $\overline{r^2}$ for a Gaussian chain considering only chain conformations for which $r > 1/2b$.

8. Calculate $\bar{r}$, the average end-to-end distance for a Gaussian chain.

9. Suppose that a Gaussian chain is attached to the ends of a rectangular box as shown.

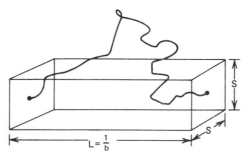

Further, suppose that the length of the box L is equal to the most probable end-to-end separation of the chain $(1/b)$ and that the ends of the box are square, having cross-sectional area S^2.

(a)   Derive an expression relating P, the pressure on the ends of the box, to the temperature.

(b)   What temperature is necessary to generate a pressure of 100 atm if $L = 100$ Å and $S^2 = 10$Å^2?

REFERENCES

1.   H. Eyring, *Phys. Rev.*, **39**, 746 (1932).

2.   F. T. Wall, *J. Chem. Phys.*, **11**, 67 (1943).

3.   P. Debye, *J. Chem. Phys.*, **14**, 636 (1946).

4.   B. H. Zimm and W. H. Stockmayer, *J. Chem. Phys.*, **17**, 1301 (1949).

5.   Lord Raleigh, *Philos. Mag.*, **37**, 321 (1919).

6.   E. Guth and H. Mark, *Monatsh.*, **65**, 93 (1934).

7.   W. Kuhn, *Kolloid-Z.*, **68**, 2 (1934).

8.   S. Chandrasekhar, *Rev. Mod. Phys.*, **15**, 3 (1939).

9.   M. V. Volkenshtein, *Configurational Statistics of Polymeric Chains*, Interscience, New York, 1963.

10.   W. Kuhn, *Kolloid-Z.*, **76**, 256 (1936); *Kolloid-Z.*, **87**, 3 (1939).

11.   T. M. Birshtein and O. B. Ptitsyn, *Conformations of Macromolecules*, Interscience, New York, 1966.

12.   F. Bueche, *Physical Properties of Polymers*, Interscience, New York, 1962, Chapter 1.

13.   P. J. Flory, *Principles of Polymer Chemistry*, Cornell University Press, Ithaca, N.Y., 1950, Chapter 10.

14.   L. R. G. Treloar, *Physics of Rubber Elasticity*, 2nd ed., Oxford University Press, London, 1958, Chapter 3.

6

Rubber Elasticity

The high elasticity of rubberlike materials is certainly their most striking characteristic. A crosslinked rubber strip, extended to several times its original length when released, will return to that original length exhibiting little or no permanent deformation as a result of the extension. This is, of course, in marked constrast to the behavior of crystalline solids and glasses, which cannot normally be extended to more than a very small fraction of their original length without undergoing fracture. It is also in contrast to the behavior of ductile materials such as some metals, which can undergo large deformations without fracture but which do not return to their original length upon removal of the deforming stress. The elastic response of rubbers has been the subject of a great deal of study by many investigators because of its very great technological importance as well as its intrinsic scientific interest. Starting from one material, namely natural rubber, the development of polymerization techniques has resulted in a host of substances that may properly be called rubbers, and a giant synthetic-rubber industry has developed to exploit them commercially.

In this chapter, we first discuss the thermodynamics of rubber elasticity. The classical thermodynamic approach, as is well known, is only concerned with the macroscopic behavior of the material under investigation and has nothing to do with its molecular structure. The latter belongs to the realm of statistical mechanics, which is the subject of the second section. We next examine the relative significance of energy and entropy in rubber elasticity. A phenomenological treatment then follows, which provides a fuller description of the elasticity at large deformations. Finally, we discuss various effects on rubber elasticity, such as degree of crosslinking, swelling, fillers, and crystallinity.

102

A. THERMODYNAMIC TREATMENT

The experimental techniques for investigating the thermodynamic behavior of rubbers can be quite numerous and varied. It is possible to investigate the rubber's response to various kinds of deformations, such as shear, tension, and compression, and it is also possible to vary temperature, pressure, and volume. Conceptually, one of the easiest experiments to perform is to extend a rubber strip to some fixed length and then measure the restoring stress exerted by the strip as a function of temperature at constant (atmospheric) pressure. It turns out that such a simple experiment embodies the fundamental thermodynamic principles of rubber elasticity and thus serves to provide a point of departure for the ensuing development. Since we shall treat rubber elasticity by *equilibrium* thermodynamics, our measurements must be done in accordance with this condition. Experimentally, equilibrium is attained after stretching by allowing stress relaxation (see Chapter 3) to proceed for a long time until constant modulus is obtained. Achievement of equilibrium can be evidenced by the reversible changes in elastic force upon changing the temperature.

Figure 6-1 illustrates the equilibrium stress–temperature behavior of natural rubber at various extension ratios.[1] The curves are linear over considerable temperature ranges with negative slopes at low degrees of elongation and positive slopes at higher degrees of elongation. This behavior is not confined to natural rubber but is general for all materials that are called rubbery or rubberlike.

The analysis of the experimental results begins with the combined first and second laws of thermodynamics, applicable to reversible processes. As shown in the preceding chapter, it reads

$$dU = T\,dS - dW \tag{6-1}$$

In equation (6-1) the increment of work, dW, refers to all of the work performed by the system (i.e., electrical, mechanical, pressure–volume, chemical, etc.). The development of thermodynamics given in most physical chemistry texts is confined to gases where dW becomes simply pressure–volume work, $P\,dV$. In the case of a rubber deformed by an amount dL in tension and exerting a restoring force f, the mechanical work performed on the system to accomplish the deformation, namely $f\,dL$, must also be included in dW. Thus, for a rubber strained uniaxially in tension,

$$dW = P\,dV - f\,dL \tag{6-2}$$

In the measurements discussed above, P is the atmospheric pressure and

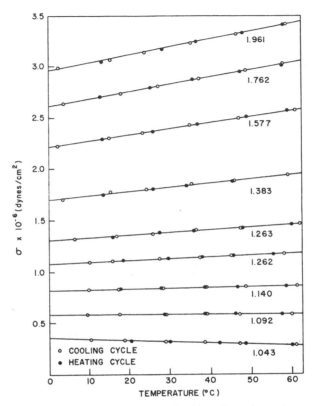

Figure 6-1. Stress–temperature curves of natural rubber. Extension ratios are indicated on the right of the figure. After M. Shen, D. A. McQuarrie, and J. L. Jackson, *J. Appl. Phys.*, **38**, 791 (1967), by permission of the American Institute of Physics.

dV is the volume dilation accompanying the elongation of the rubber. Generally, $P\,dV$ is much smaller than $f\,dL$ and may sometimes be neglected in comparison to $f\,dL$. However, for completeness the treatment following will include pressure–volume work.

Equation (6-1) now becomes

$$dU = T\,dS - P\,dV + f\,dL \tag{6-3}$$

Since the experiments discussed were performed under constant pressure conditions, it is natural to use the thermodynamic entity H, the enthalpy. H is defined as

$$H = U + PV \tag{6-4}$$

Differentiating equation (6-4) at constant pressure results in

$$dH = dU + P\,dV \tag{6-5}$$

Substituting for dU from equation (6-3) we obtain

$$dH = T\,dS + f\,dL \tag{6-6}$$

Thus, the restoring force exerted by the rubber when it has undergone a deformation by the amount dL at constant temperature and pressure is:

$$f = \left(\frac{\partial H}{\partial L}\right)_{T,P} - T\left(\frac{\partial S}{\partial L}\right)_{T,P} \tag{6-7}$$

Equation (6-7) shows that the restoring force originates in the enthalpy and entropy changes that occur in the elastomer as a result of the deformation that it undergoes.

The change in slope of the stress–temperature curve from negative values at low degrees of elongation to positive values at high degrees of elongation is known as the thermoelastic inversion phenomenon (Figure 6-1). In order to see why this should occur, it is necessary to derive a thermodynamic expression for the slope of this curve, that is, an expression for $(\partial f/\partial T)_{P,L}$. This can be done as follows. The Gibbs free energy is defined by

$$dF = -S\,dT + V\,dP + f\,dL \tag{6-8}$$

One of Maxwell's relations states that

$$\left(\frac{\partial S}{\partial L}\right)_{T,P} = -\left(\frac{\partial f}{\partial T}\right)_{P,L} \tag{6-9}$$

Substitution for $(\partial S/\partial L)_{T,P}$ in equation (6-7) yields

$$f = \left(\frac{\partial H}{\partial L}\right)_{T,P} + T\left(\frac{\partial f}{\partial T}\right)_{P,L} \tag{6-10}$$

Which, upon rearrangement, leads to the desired result:

$$\left(\frac{\partial f}{\partial T}\right)_{P,L} = \frac{f - (\partial H/\partial L)_{T,P}}{T} \tag{6-11}$$

Thus, in order for the slope of the f versus T curves to be negative, it must be that $(\partial H/\partial L)_{T,P} > f$, while in order for the slope to be positive, it must be that $f > (\partial H/\partial L)_{T,P}$. The experimental results show that the former inequality holds at low elongations and the latter inequality holds at high elongations. At sufficiently high elongations it is possible to neglect $(\partial H/\partial L)_{T,P}$ in comparison with f and as a result, under these conditions

$$f = T\left(\frac{\partial f}{\partial T}\right)_{P,L} = -T\left(\frac{\partial S}{\partial L}\right)_{T,P} \qquad (6\text{-}12)$$

Thus f becomes directly proportional to the absolute temperature, or the elastic response of the rubber is entirely governed by the decrease in entropy which it undergoes upon extension.

Although equation (6-12) adequately describes the behavior of an elastomer at high extensions ($>10\%$ for natural rubber), it is nevertheless true that the coefficient $(\partial H/\partial L)_{T,P}$ has a finite value and cannot be neglected in an exact treatment of rubber elasticity. [It should be stated that equation (6-12) is inadequate to describe the behavior of most elastomers at very high extensions. This is because many elastomers undergo crystallization at sufficiently high extensions and, when this occurs, the term $(\partial H/\partial L)_{T,P}$ once again becomes important and may actually outweigh the term $-T(\partial S/\partial L)_{T,P}$.] In order to explore the origins of the coefficient $(\partial H/\partial L)_{T,P}$ we return to equation (6-5). Differentiation of equation (6-5) at constant temperature yields:

$$\left(\frac{\partial H}{\partial L}\right)_{T,P} = \left(\frac{\partial U}{\partial L}\right)_{T,P} + P\left(\frac{\partial V}{\partial L}\right)_{T,P} \qquad (6\text{-}13)$$

We wish to separate effects arising from volume changes and effects arising from deformations of the material. To do this, we proceed as follows. Expressing U as a function of length and volume,

$$dU = \left(\frac{dU}{dV}\right)_{L,T} dV + \left(\frac{\partial U}{\partial L}\right)_{V,T} dL \qquad (6\text{-}14)$$

From equation (6-14):

$$\left(\frac{\partial U}{\partial L}\right)_{T,P} = \left(\frac{\partial U}{\partial V}\right)_{L,T}\left(\frac{\partial V}{\partial L}\right)_{T,P} + \left(\frac{\partial U}{\partial L}\right)_{V,T} \qquad (6\text{-}15)$$

which, upon substitution into equation (6-13) and rearrangement, gives

$$\left(\frac{\partial H}{\partial L}\right)_{T,P} = \left(\frac{\partial U}{\partial L}\right)_{T,V} + \left[\left(\frac{\partial U}{\partial V}\right)_{T,L} + P\right]\left(\frac{\partial V}{\partial L}\right)_{T,P} \qquad (6\text{-}16)$$

Equation (6-16) shows that the coefficient $(\partial H/\partial L)_{T,P}$ consists of a part arising from internal energy changes occurring during a change of length at constant volume and a volume-dependent part. Although the thermodynamic development is not concerned with the molecular nature of the rubber, the coefficient $(\partial U/\partial L)_{T,V}$ can be related to intramolecular energy effects. This is so because of the constant volume condition. Real rubbers usually have nonzero values for this coefficient.

The coefficient $(\partial U/\partial V)_{T,L}$ may be transformed into experimentally accessible quantities as follows. The Helmholtz free energy, dA, is given by

$$dA = -S\,dT - P\,dV + f\,dL \qquad (6\text{-}17)$$

From this is follows that

$$\left(\frac{\partial S}{\partial V}\right)_{T,L} = \left(\frac{\partial P}{\partial T}\right)_{V,L} \qquad (6\text{-}18)$$

Returning to equation (6-3) and differentiating with respect to V at constant T and L:

$$\left(\frac{\partial U}{\partial V}\right)_{T,L} = T\left(\frac{\partial S}{\partial V}\right)_{T,L} - P \qquad (6\text{-}19)$$

or, making use of equation (6-18):

$$\left(\frac{\partial U}{\partial V}\right)_{T,L} = T\left(\frac{\partial P}{\partial T}\right)_{V,L} - P \qquad (6\text{-}20)$$

By a standard technique of partial differentiation,

$$\left(\frac{\partial P}{\partial T}\right)_{V,L} = -\left(\frac{\partial P}{\partial V}\right)_{T,L}\left(\frac{\partial V}{\partial T}\right)_{P,L} \qquad (6\text{-}21)$$

The cubical coefficient of thermal expansion is defined by

$$\alpha = \frac{1}{V}\left(\frac{\partial V}{\partial T}\right)_{P,L} \qquad (6\text{-}22)$$

and the coefficient of isothermal compressibility by

$$\beta = -\frac{1}{V}\left(\frac{\partial V}{\partial P}\right)_{T,L} \qquad (6\text{-}23)$$

(Note that, as defined above, both α and β are functions of length inasmuch as the volume of the elastomer undergoes changes on extension.) Using these definitions, we have

$$\left(\frac{\partial P}{\partial T}\right)_{V,L} = \frac{\alpha}{\beta} \tag{6-24}$$

and

$$\left(\frac{\partial U}{\partial V}\right)_{T,L} = T\frac{\alpha}{\beta} - P \tag{6-25}$$

Inserting equation (6-25) into equation (6-16):

$$\left(\frac{\partial H}{\partial L}\right)_{T,P} = \left(\frac{\partial U}{\partial L}\right)_{T,V} + T\frac{\alpha}{\beta}\left(\frac{\partial V}{\partial L}\right)_{T,P} \tag{6-26}$$

The quantity $(\partial H/\partial L)_{T,P}$ is directly accessible from experimental f versus T curves as the intercept at $T=0$ of the tangent to the experimental curve at any desired temperature [see equation (6-10)]. Its measurement thus involves no difficulty for any elastomer provided equilibrium constant-pressure stress–temperature data are available for that elastomer. In principle, the second term on the right side of equation (6-26) is also directly measurable, but precise values turn out to be very difficult to obtain in practice. Measurements of α and β can be carried out with relative ease, but the coefficient $(\partial V/\partial L)_{T,P}$ is very small for most rubbers and usually depends on extension; thus great experimental skill is required in order to obtain a measurement of any accuracy.[2,3]

By combining equation (6-26) with equation (6-10), the energetic component of the elastic force f_e becomes

$$f_e = \left(\frac{\partial U}{\partial L}\right)_{T,V}$$

$$= f - T\left(\frac{\partial f}{\partial T}\right)_{P,L} - \frac{T\alpha}{\beta}\left(\frac{\partial V}{\partial L}\right)_{T,P} \tag{6-27}$$

Thus, from equation (6-27) one should be able to compute the relative importance of contributions from energy and entropy to rubber elasticity. Once again, the unavailability of accurate values of the partial $(\partial V/\partial L)_{T,P}$ renders this separation difficult at best.

In general, direct measurements[4] have not been fruitful and various approximate methods have been devised. The latter are discussed in detail in Section B2 of this chapter.

B. STATISTICAL TREATMENT

A rubberlike solid is unique in that its physical properties resemble those of solids, liquids, and gases in various respects. It is solidlike in that it maintains dimensional stability, and its elastic response at small strains ($<5\%$) is essentially Hookean. It behaves like a liquid because its coefficient of thermal expansion and isothermal compressiblility are of the same order of magnitude as those of liquids. The implication of this is that the intermolecular forces in a rubber are similar to those in liquids. It resembles gases in the sense that the stress in a deformed rubber increases with increasing temperature, much as the pressure in a compressed gas increases with increasing temperature. This gaslike behavior was, in fact, what first provided the hint that rubbery stresses are entropic in origin.

1. Derivation

The molecular model for the ideal gas is a collection of point masses in ceaseless, random, thermal motion, the motion of any two of the point masses being completely uncorrelated with one another. The counterpart to this in the case of the ideal rubber is a collection of volumeless, long, flexible chains that are continually undergoing conformational rearrangements due to thermal motion but with all conformations equally accessible and without energy differences among conformations. (That is, all conformations are isoenergetic.) It is assumed that some chain conformations become inaccessible upon stretching, leading to a decrease in entropy and hence to the restoring stress exerted by the rubber.

In order for a rubber to exhibit an equilibrium elastic stress, it is necessary for the collection of linear polymer chains to be tied together into an infinite network. Otherwise the Brownian motions of the macromolecules will cause them to move past each other, thus exhibiting flow (Chapter 3). Crosslinking reactions are many and varied. It suffices for our purposes to consider a crosslink to be a permanent tie-point between two chains (Figure 6-2).

To compare the ideal rubber with the ideal gas on a more quantitative basis, we note from equation (6-3) that

$$f = \left(\frac{\partial U}{\partial L}\right)_{T,V} - T\left(\frac{\partial S}{\partial L}\right)_{T,V} \qquad (6\text{-}28)$$

Equation (6-28) resembles the relation

$$-P = \left(\frac{\partial U}{\partial V}\right)_T - T\left(\frac{\partial S}{\partial V}\right)_T \qquad (6\text{-}29)$$

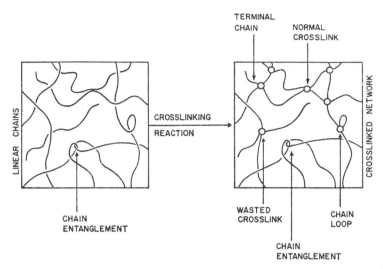

Figure 6-2. Schematic diagram of an ensemble of linear polymer chains being crosslinked into an infinite network.

applicable to gases. For ideal gases the internal energy is independent of volume, $(\partial U/\partial V)_T = 0$, and the entropy has two components: one is associated with the heat capacity of the gas, but independent of volume; the other is related to the configurational entropy of the system, and thus a function of the volume. By analogy, the ideal rubber may be looked at in the same way. Its internal energy is independent of elongation as required by the model, and thus $(\partial U/\partial L)_{T,V} = 0$ and the stress can be attributed to the configurational entropy alone.

The term configurational entropy has been applied historically to the volume-dependent portion of the entropy in an ideal gas; by analogy, the elongation-dependent entropy in the ideal rubber was also referred to as configurational entropy. A difficulty arises with this terminology in rubbers. Rubber molecules are capable of geometrical isomerization, such as cis and trans isomers as well as stereoisomerism, which results in, for example, isotactic and syndiotactic forms. These various isomers are known as configurations and cannot be transformed from one into another without breaking chemical bonds. Thus, upon stretching, no configurational changes are possible and there is no configurational contribution to the entropy. What does take place are rotations about single bonds in the chain backbone. It is these conformational changes that give rise to the entropy decrease upon stretching. The conformational statistics of a single chain were the subject of the last chapter. Thus conformational entropy changes upon stretching are the major sources of rubber elasticity, and we shall refer to the entropy involved as conformational entropy in the discussion that follows.

In the formulation of the statistical theory of rubber elasticity,[5-11] the following simplifying assumptions are made:

1. The internal energy of the system is independent of the conformations of the individual chains.

2. An individual network chain is freely joined and volumeless, that is, it obeys Gaussian statistics (Chapter 5, Section B).

3. The total number of conformations of an isotropic network of such Gaussian chains is the product of the number of conformations of the individual network chains.

4. Crosslink junctions in the network are fixed at their mean positions. Upon deformation, these junctions transform affinely, that is, in the same ratio as the macroscopic deformation ratio of the rubber sample.

Now, from equation (6-17):

$$f = \left(\frac{\partial A}{\partial L} \right)_{T,V} \tag{6-30}$$

But by definition,

$$A = U - TS \tag{6-31}$$

Because of assumption 1, however, we need not find an explicit expression for U but can concentrate on the entropy expression. For this we shall again invoke the Boltzmann relation (equation (5-53)), as we did for the isolated chain:

$$S = k \ln \Omega \tag{6-32}$$

where Ω is the total number of conformations available to the rubber network. According to assumption 2, the number of conformations available to the ith individual chain is given by the Gaussian distribution function [equation (5-29)]:

$$\omega(x_i, y_i, z_i) = \left(\frac{b}{\pi^{1/2}} \right)^3 \exp\left[-b^2(x_i^2 + y_i^2 + z_i^2) \right] \tag{6-33}$$

Equation (6-33) refers to a chain having an end-to-end vectorial distance $\mathbf{r}_i$

with one end at the coordinates (x_i, y_i, z_i) in the unstrained state, the other end being at the origin of the Cartesian coordinate system (Figure 5-2). Following assumption 3, the total number of conformations available to a network of N such chains is

$$\Omega = \prod_{i=1}^{N} \omega(\mathbf{r}_i) \tag{6-34}$$

and the conformational entropy of the undeformed network is just

$$S_u = 3k \ln \frac{b}{\pi^{1/2}} - k \sum_{i=1}^{N} b^2 (x_i^2 + y_i^2 + z_i^2) \tag{6-35}$$

or

$$A_u = A_0 + kT \sum_{i=1}^{N} b^2 (x_i^2 + y_i^2 + z_i^2) \tag{6-36}$$

where A_0 is that portion of the Helmholtz free energy that is not related to conformational entropy changes.

In the strained state, the chain is deformed to $\mathbf{r}_i'$ with the chain end now at coordinates (x_i', y_i', z_i'). To relate the microscopic strain of the chains to the macroscopic strain of the rubber sample, we assume the deformation to be affine (assumption 4). Consider a unit cube of an isotropic rubber sample (Figure 6-3a). In the general case of a pure homogeneous strain, the cube is transformed into a rectangular parallelepiped (Figure 6-3b). The dimensions of the parallelpiped are λ_1, λ_2, and λ_3 in the three principal axes, where the λ's are called the principal extension ratios. Choosing the coordinate axes for the chain to coincide with the principal axes of strain for the sample, then

$$x_i' = \lambda_1 x_i, \qquad y_i' = \lambda_2 y_i, \qquad z_i' = \lambda_3 z_i \tag{6-37}$$

The Helmholtz free energy of the deformed network can thus be written as

$$A_d = A_0 + kT \sum_{i=1}^{N} b^2 (\lambda_1^2 x_i^2 + \lambda_2^2 y_i^2 + \lambda_3^2 z_i^2) \tag{6-38}$$

The total change in free energy of the rubber network due to the deformation is simply the difference between equations (6-36) and (6-38):

$$\Delta A = kT \left[(\lambda_1^2 - 1) \sum_{i=1}^{N} b^2 x_i^2 + (\lambda_2^2 - 1) \sum_{i=1}^{N} b^2 y_i^2 + (\lambda_3^2 - 1) \sum_{i=1}^{N} b^2 z_i^2 \right] \tag{6-39}$$

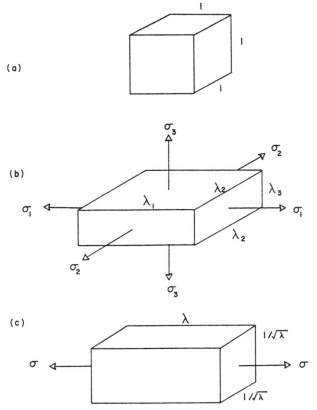

Figure 6-3. A unit cube of rubber: (a) in the unstrained state; (b) in homogeneous strained state; (c) under uniaxial extension.

We recall that by definition:

$$r_i^2 = x_i^2 + y_i^2 + z_i^2 \tag{6-40}$$

For a random, isotropic network all directions are equally probable; then

$$x_i^2 = y_i^2 = z_i^2 = \tfrac{1}{3} r_i^2 \tag{6-41}$$

Equation (6-39) now becomes

$$\Delta A = \frac{1}{3} kT \sum_{i=1}^{N} b^2 r_i^2 (\lambda_1^2 + \lambda_2^2 + \lambda_3^2 - 3)$$

$$= \frac{1}{3} kNT \overline{b^2 r^2} (\lambda_1^2 + \lambda_2^2 + \lambda_3^2 - 3) \tag{6-42}$$

where we have written $\overline{b^2r^2}=\sum_{i=1}^{N}b^2r_i^2/N$. In Chapter 5, Section B, we defined b^2 for an unstrained freely orienting random chain as

$$b^2 = \frac{3}{2\overline{r_f^2}} \tag{6-43}$$

If the network chains in the unstrained state have the same distribution of conformations as an ensemble of free chains, then $\overline{b^2r^2}=3/2$. However, in real networks this condition may not be met. For instance, some of the chains may already be partially strained during the crosslinking process. The details of the crosslinking process, that is, whether it is carried out in solution or in bulk, may also affect the state of the network. It is therefore more general to write the following:

$$\overline{b^2r^2} = \overline{b^2} \sum_{i=1}^{N} \frac{r_i^2}{N} = \frac{3}{2}\frac{\overline{r_0^2}}{\overline{r_f^2}} \tag{6-44}$$

where $\overline{b^2}$ is averaged over all the free chains, that is $\overline{b^2}=3/2\overline{r_f^2}$. Here $\overline{r_0^2}=\sum r_i^2/N$ refers to the mean square end-to-end distance of the chain in the network, and $\overline{r_f^2}$ to the mean square end-to-end distance of the isolated chain. Substitution of equation (6-44) into equation (6-42) yields

$$\Delta A = \frac{T}{2}Nk\frac{\overline{r_0^2}}{\overline{r_f^2}}(\lambda_1^2+\lambda_2^2+\lambda_3^2-3) \tag{6-45}$$

The parameter $\overline{r_0^2}/\overline{r_f^2}$, sometimes referred to as the front factor, can be regarded as the average deviation of the network chains from the dimensions they would assume if they were isolated and free from all constraints. For an ideal rubber network, the front factor is unity.

The treatment of rubber elasticity presented above represents one possible extreme of behavior. The assumption that the crosslink points in the network are fixed at their mean positions and that the crosslink points deform affinely gives rise to this extreme. In real polymer networks, each crosslink point finds itself in the neighborhood of many other crosslink points. This can be verified by estimating the order of magnitude of the concentration of crosslinks and then calculating the number of crosslink points that would be found within some reasonable distance (perhaps 20 Å) of any given crosslink point. Upon deformation, the affine assumption insists that all of these crosslinks remain in the neighborhood of the particular crosslink point under consideration and, moreover, that their relative positions are fixed.

In the consideration of real networks, it is clear that this assumption is overly restrictive. For example, consider a particular chain that happens to be reasonably extended in the unstrained network state. In the neighborhood of a crosslink point for this chain, we will find many other crosslink points for other chains. At least some of these other chains are expected to be in less extended configurations than the chain under consideration. Upon deformation, the tendency for the already extended chain to further elongate would be expected to be less than the tendency for the more relaxed chains to elongate. This being the case, the positions of crosslink points would be expected to move past one another in a manner not strictly defined by the affine deformation.

The other extreme of behavior involves the "phantom chain" approximation. Here, it is assumed that the individual chains and crosslink points may pass through one another as if they had no material existence, that is, they may act like phantom chains. In this approximation, the *mean* position of crosslink points in the deformed network is consistent with the affine transformation, but *fluctuations* of the crosslink points are allowed about their mean positions and these fluctuations are not affected by the state of strain in the network. Under these conditions, the distribution function characterizing the position of crosslink points in the deformed network cannot be simply related to the corresponding distribution function in the undeformed network via an affine transformation. In this approximation, the crosslink points are able to readjust, moving through one another, to attain the state of lowest free energy subject to the deformed dimensions of the network.

For an ideal network with tetrafunctional crosslink points, it can be shown for the phantom chain approximation that[12]

$$\Delta A = \frac{T}{4} Nk \left(\lambda_1^2 + \lambda_2^2 + \lambda_3^2 - 3 \right) \tag{6-46}$$

It is clear that the Helmholtz free-energy change upon deformation predicted by equation (6-46) is just one-half as large as that given in equation (6-45) based on the affine deformation of crosslink points. Real networks probably exhibit behavior that is between these two extremes. Not all crosslink points define affinely but, on the other hand, steric interactions are strong enough to prevent the phantom approximation from being completely realistic.

From this point forward, we continue with our development of rubber elasticity theory using equation (6-45) as a starting point. We must remember, however, that this equation is based on a limiting approximation and that the behavior of real networks may not be quantitatively consistent with this expression.

Suppose our unit cube has volume V_0 and length L_0. After a uniaxial extension ($\lambda_1 = L/L_0$) is applied, the length in the direction of stretch is L and the volume dilates to V. We have mentioned previously that the strain-induced volume dilation for unfilled rubbers is very small, of the order of magnitude of 10^{-4}. Nevertheless, the deformation process is not a volume-preserving one, as required by equation (6-30). The device that is generally employed to circumvent this difficulty is to redefine the reference state. A hypothetical hydrostatic pressure is imagined to have been applied to the sample so that its volume in the unstretched state is also V. The initial length is no longer L_0, but is

$$L' = L_0\left(\frac{V}{V_0}\right)^{1/3}\qquad(6\text{-}47)$$

and the extension ratio in the direction of the uniaxial stretch becomes

$$\alpha^* = \frac{L}{L'}$$

$$\alpha_1^* = \lambda_1\left(\frac{V_0}{V}\right)^{1/3}\qquad(6\text{-}48)$$

The average chain dimensions become $\overline{r^2} = \overline{r_0^2}(V/V_0)^{2/3}$. Now the condition of incompressibility requires that

$$\alpha_1^*\alpha_2^*\alpha_3^* = 1\qquad(6\text{-}49a)$$

while previously

$$\lambda_1\lambda_2\lambda_3 = \frac{V}{V_0}\qquad(6\text{-}49b)$$

Since the network is isotropic, the contractions along the two lateral axes are equal and

$$\lambda_1 = \alpha_1^*\left(\frac{V}{V_0}\right)^{1/3}$$

$$\lambda_2 = \lambda_3 = \alpha_1^{*-1/2}\left(\frac{V}{V_0}\right)^{1/3}\qquad(6\text{-}50)$$

The uniaxial extension of the unit cube is illustrated in Figure 6-3c. Inserting equation (6-50) into equation (6-45), the change in Helmholtz free energy upon deformation becomes

$$\Delta A = \frac{T}{2} Nk \frac{\overline{r_0^2}}{r_f^2} \frac{\alpha^{*2}+2}{\alpha^*-3} \qquad (6\text{-}51)$$

for simple uniaxial extension. Since

$$\left(\frac{\partial A}{\partial L}\right)_{T,V} = \left(\frac{\partial A}{\partial \alpha^*}\right)_{T,V} \left(\frac{\partial \alpha^*}{\partial L}\right)_{T,V} \qquad (6\text{-}52a)$$

$$f = \frac{1}{L'}\left(\frac{\partial A}{\partial \alpha^*}\right)_{T,V} \qquad (6\text{-}52b)$$

Performing the indicated differentiation on equation (6-51), we obtain the equation of state for rubber elasticity:

$$f = \frac{NkT}{L'} \frac{\overline{r_0^2}}{r_f^2} \left[\alpha^* - \frac{1}{\alpha^{*2}}\right]\left(\frac{V}{V_0}\right)^{2/3} \qquad (6\text{-}53)$$

In equation (6-53) f is the *total* elastic restoring force exerted by the sample. For many purposes it is more convenient to deal with expressions relating the stress to the deformation rather than the total force as in equation (6-53). For this purpose we define a stress: $\sigma = f/A_0$, where A_0 is the cross-sectional area of the undeformed sample. We further define $N_0 = N/V_0$ as the number of network chains per unit volume of the undeformed sample, where $V_0 = L_0 A_0$. We shall also express N_0 in terms of the number of *moles* of network chains. With the aid of equation (6-48), we can rewrite equation (6-53) as

$$\sigma = N_0 RT \frac{\overline{r_0^2}}{r_f^2}\left(\lambda - \frac{V}{V_0\lambda^2}\right) \qquad (6\text{-}54)$$

where R is the ideal gas constant ($R = N_{Av}k$, N_{Av} being the Avogadro's number). The ratio V/V_0 is very nearly unity; thus it is often adequate to write equation (6-54) as

$$\sigma = N_0 RT \frac{\overline{r_0^2}}{r_f^2}\left(\lambda - \frac{1}{\lambda^2}\right) \qquad (6\text{-}55)$$

The difference between equations (6-54) and (6-55) is numerically trivial, although conceptually important.

The extension ratio may be written

$$\lambda = 1 + \varepsilon \tag{6-56}$$

where ε is the tensile strain, $\Delta L / L_0$. By the binomial expansion

$$\lambda^{-2} = (1 + \varepsilon)^{-2} = 1 - 2\varepsilon + \cdots \tag{6-57}$$

For very small strains, higher-order terms can be neglected and equation (6-55) may be recast as

$$E_0 = \frac{\sigma}{\varepsilon} = 3N_0 RT \frac{\overline{r_0^2}}{\overline{r_f^2}} \tag{6-58}$$

where E_0 is the time-independent tensile modulus. Since for crosslinked rubbers the tensile modulus is three times the shear modulus G_0 (Chapter 2, Section A), then

$$G_0 = N_0 RT \frac{\overline{r_0^2}}{\overline{r_f^2}} \tag{6-59}$$

Thus the equation of state can be recast as

$$f = G_0 A_0 \left(\lambda - \frac{1}{\lambda^2} \right) \tag{6-60a}$$

or

$$\sigma = G_0 \left(\lambda - \frac{1}{\lambda^2} \right) \tag{6-60b}$$

where G_0 is determined by the initial slope in the stress–strain curve and defined by equation (6-59).

The equation of state for rubber elasticity, embodied by any of equations (6-53) through (6-59), is important not only because it is historically the first quantitative treatment of molecular theories for rubbers but also because it laid a conceptual foundation for theories for the physical properties of polymers in general. In later chapters we shall have occasion to discuss some of these in detail. Perhaps the single most significant

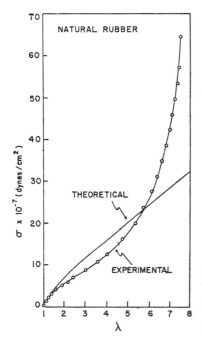

Figure 6-4. Stress-strain curve for natural rubber. The theoretical curve was calculated from equation (6-60) with $G_0 = 4 \times 10^6$ dynes/cm². After L. R. G. Treloa, *Trans. Faraday Soc.*, **40**, 59 (1944), by permission of the Faraday Society.

contribution is its recognition of the role of entropy in polymers, in contradistinction to the predominant role of energy in "ordinary" solids. The equation predicts that the elastic force is directly proportional to temperature and to the total number of chains in the network. Both of these are experimentally observed. The strain dependence of the elastic force is clearly not Hookean, as it is for a single polymer chain [equation (5-56)]. Figure 6-4 compares the theory with the experimental stress–strain curve.[13] The agreement is good for strains below 50% or $\lambda < 1.5$, but is poor at high extensions. At these high strains the network chains approach their limiting extensions, and the Gaussian assumption is no longer valid. Another complicating factor is the onset of strain-induced crystallization, which we discuss further in Section D. Even though the Gaussian theory is only valid at relatively low strains, it is extremely valuable in providing a molecular interpretation for rubber elasticity.

2. Energy Contribution

In our statistical treatment of an ideal rubber, we have assumed that the elastic force is entirely attributable to the conformational entropy of deformation, energy effects being neglected. That the theory reproduces the essential features of the elasticity of real rubbers attests to the basic soundness of this assumption. On the other hand, we know that in real rubbers such energy effects cannot be entirely absent, and deviations from the ideal rubber model may be expected to occur. Let us now examine in greater detail the extent to which the neglect of energy effects is justified.

We can rewrite equation (6-28):

$$f = f_e + f_s \tag{6-61}$$

with

$$f_e = \left(\frac{\partial U}{\partial L}\right)_{T,V} \tag{6-62}$$

$$f_s = -T\left(\frac{\partial S}{\partial L}\right)_{T,V} \tag{6-63}$$

Since we can experimentally determine f and obtain f_s from the thermodynamic identity [Maxwell relationship resulting from equation (6-17)],

$$\left(\frac{\partial S}{\partial L}\right)_{T,V} = -\left(\frac{\partial f}{\partial T}\right)_{V,L} \tag{6-64}$$

We can find f_e by

$$f_e = f - T\left(\frac{\partial f}{\partial T}\right)_{V,L} \tag{6-55}$$

As discussed in the previous section, the condition of constant volume is difficult to achieve experimentally. Similarly, the alternative expression, equation (6-28), also contains terms that are not accurately measurable. We can, however, take advantage of the exact expression of the statistical theory:

$$f = G_0 A_0\left(\lambda - \frac{V}{V_0\lambda^2}\right) \tag{6-66}$$

where G_0 is defined by equation (6-59). The shear modulus G_0 is a material constant, which is invariant whether the experimental condition is constant volume or constant pressure. We can therefore differentiate equation (6-66) with respect to T, keeping V and L constant. (Note, however, that V_0 and

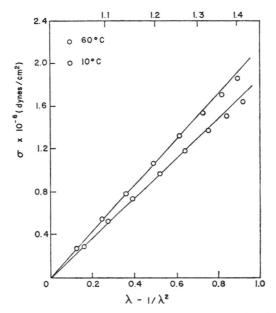

Figure 6-5. Determination of shear moduli for natural rubber at 10°C and 60°C by plotting σ against $\lambda - 1/\lambda^2$. After M. Shen and P. J. Blatz *J. Appl. Phys.*, **39**, 4937 (1968), by the permission of the American Institute of Physics.

L_0 are not constant.) The result is

$$\left(\frac{\partial f}{\partial T}\right)_{V,L} = \frac{f\,dG_0}{G_0\,dT} + \frac{\alpha f}{3} \tag{6-67}$$

Combining equations (6-67) and (6-65), we obtain:

$$f_e = f - f\frac{d\ln G_0}{d\ln T} - \frac{\alpha T f}{3} \tag{6-68}$$

The advantage of equation (6-68) is that experimental errors in stress–strain data at various temperatures are averaged out in plotting f against $(\lambda - \lambda^{-2})$, the slope of which is the shear modulus (Figure 6-5). Knowing shear moduli as a function of temperature, values of f_e can be readily calculated.

For the sake of comparison, it is more meaningful to use the relative energy contribution f_e/f. Thus equation (6-68) can be rewritten as:

$$\frac{f_e}{f} = 1 - \frac{d\ln G_0}{d\ln T} - \frac{\alpha T}{3} \tag{6-69}$$

Table 6-1. Thermoelastic Data of Selected Elastomers (Reference temperature: 30°C)

Polymer	$\dfrac{\alpha}{3} \times 10^4$ (cm/cm/°C)	$\dfrac{d \ln G_0}{dT} \times 10^3$ (°C^{-1})	f_e/f	Ref.
Poly(ethylene-co-propylene) (EPR)	2.5	2.9	0.04	[b]
Poly(tetrafluoroethylene-co-perfluoropropylene) (Viton A)	3.0	2.8	0.05	[b]
Poly(butadiene-co-styrene) (SBR)	2.9	3.4	−0.12	[b]
Poly(butadiene-co-acrylonitrile) (Hycar)	2.6	2.9	0.03	[b]
Poly(cis-1,4-butadiene) (Budene)	2.1	2.8	0.10	[b]
Poly(cis-isoprene) (natural rubber)	2.2	2.5	0.18	[c]
Poly(2-hydroxypropyl acrylate)	2.8	3.6	−0.53[a]	[d]
Poly(isobutylacrylate)	2.4	3.4	−0.42[a]	[d]
Poly(isobutylmethacrylate)	2.2	2.3	0.02[a]	[d]

[a] Reference temperature: 120° C.
[b] E. H. Cirlin, H. M. Gebhard, and M. Shen, *J. Macromol. Sci.,* Part A, **5**, 981 (1971).
[c] M. Shen, *Macromolecules,* **2**, 358 (1969).
[d] M. Shen, E. H. Cirlin, and H. M. Gebhard, *Macromolecules,* **2**, 682 (1969).

Values of f_e/f calculated from equation (6-69) are independent of λ, as long as they are obtained within the region of strain for which the Gaussian theory is valid.

Table 6-1 shows the values of f_e/f for several rubbers. We see that in general we cannot expect the contribution of energy to rubber elasticity to be zero. Rather, a fraction of the stress is attributable to energy, the rest to entropy. However, since this fraction is a constant as a function of strain, the general *shape* of the stress–strain curve is still unaffected. Thus the neglect of energy effects in the statistical theory essentially causes the predicted curve to differ from the experimental one by a constant fraction. The shear modulus in the statistical theory expression [equation (6-60)] is normally obtained by fitting the experimental stress–strain curve rather than calculating *a priori* from equation (6-59) (neither N_0 nor $\overline{r_0^2}/\overline{r_f^2}$ can at present be determined with certainty by independent measurements). For this reason the energy component of the elastic force is often "absorbed" in the equation of state, rather than separated according to equation (6-61).

We recall that one of the assumptions used in the derivation of the statistical theory is the free energy additivity principle (assumption 3). According to this principle, the number of conformations available to a network of chains is just the product of all the conformations of the individual chains, or the entropy of the network is the sum of the entropies of individual chains in the network. In order for this assumption to be valid, it is required that chains in the network behave as if they were in free space and unaffected by the presence of other chains. This stipulation can only be satisfied if interchain interactions are absent. Thus the energy effects present in rubber elasticity must only come from intrachain interactions, such as the energy barriers hindering rotations along the polymer chain. It is reasonable to expect that real polymer chains are not isoenergetic, that is, energies of the individual chains are not constant as a function of their conformations. It follows that changes in the supply of thermal energy (changes in temperature) would produce changes in the mean chain dimensions as well. We can derive an expression for the temperature coefficient of the unperturbed dimension of the polymer chain by differentiating equation (6-59) and remembering that $N_0 = N/V_0$ and r_0^2 is proportional to $V_0^{2/3}$:

$$\frac{d \ln \overline{r_f^2}}{d \ln T} = -1 - \frac{d \ln G_0}{d \ln T} - \frac{\alpha T}{3} \qquad (6\text{-}70)$$

Inserting equation (6-70) into equation (6-69), we see that

$$\frac{d \ln \overline{r_f^2}}{d \ln T} = \frac{f_e}{f} \qquad (6\text{-}71)$$

Thus the energy effects in rubber elasticity arise from the intrachain interaction energies of the network chains.

C. PHENOMENOLOGICAL TREATMENT

The statistical theory of rubber elasticity discussed in the preceding section was arrived at through considerations of the underlying molecular structure. The equation of state was obtained directly from the Helmholtz free energy of deformation (or simply conformational entropy of deformation,

since the energy effects were assumed to be absent), which we can recast
with the aid of equations (6-45) and (6-59) as

$$A = -TS = \tfrac{1}{2}G_0(\lambda_1^2 + \lambda_2^2 + \lambda_3^2 - 3) \tag{6-72}$$

As we have already seen in Figure 6-4, the theory does not give a complete
description of the stress–strain behavior of rubberlike materials. This is
perhaps to be expected in view of the many rather drastic simplifying
assumptions that have been made in its derivation. That the theory predicts
many of the essential features of rubber elasticity is a tribute to the depth of
its insight. However, it is desirable to be able to describe mathematically
the stress–strain relations of rubbers over a larger region of strain. For this
purpose we must resort to continuum mechanics for a phenomenological
treatment of elasticity at large strains.

The phenomenological theory, as its name implies, concerns itself only
with the observed behavior of rubbers. It is not based on considerations of
the molecular structure of the polymer. The central problem here is to find
an expression for the elastic energy stored in the system, analogous to the
free energy expression in the statistical theory [equation (6-72)]. Consider
again the deformation of our unit cube in Figure. 6-3. In order to arrive at
the state of strain, a certain amount of work must be done which is stored
in the body as strain energy:

$$\overline{W}(\lambda_i) = \int_{\lambda_1=1}^{\lambda_1} \sigma_1 \, d\lambda_1 + \int_{\lambda_2=1}^{\lambda_2} \sigma_2 \, d\lambda_2 + \int_{\lambda_3=1}^{\lambda_3} \sigma_3 \, d\lambda_3 \tag{6-73}$$

where the λ's are again the principal extension ratios. This energy is a
unique function of the state of strain, and if the amount of it is known as a
function of strain, the elastic properties of the material can then be
completely defined. Although the strain energy function, expressed in terms
of the principal extension ratios, is chosen without any regard for the
molecular mechanism, it must satisfy certain logical constraints in the case
of isotropic solids. These considerations lead to the expression of $\overline{W}$ in
terms of the so-called strain invariants:[15,16]

$$\overline{W} = \overline{W}(I_i), \qquad i = 1, 2, 3 \tag{6-74}$$

where

$$I_1 = \lambda_1^2 + \lambda_2^2 + \lambda_3^2$$

$$I_2 = \lambda_1^2\lambda_2^2 + \lambda_2^2\lambda_3^2 + \lambda_3^2\lambda_1^2 \tag{6-75}$$

$$I_3 = \lambda_1^2\lambda_2^2\lambda_3^2$$

The third strain invariant is obviously

$$I_3 = \left(\frac{V_0}{V}\right)^2 \tag{6-76}$$

which is equal to unity for an incompressible material.

The most general form of the strain energy function for an isotropic material is the power series:

$$\overline{W} = \sum_{i,j,k=0}^{\infty} C_{ijk}(I_1 - 3)^i (I_2 - 3)^j (I_3 - 1)^k \tag{6-77}$$

Quantities in the parentheses are so chosen that the strain energy vanishes at zero strain. Since we cannot determine *a priori* the set of terms in equation (6-77), let us examine the lowest members of the series. For $i = 1$, $j = 0$, $k = 0$:

$$\overline{W} = C_{100}(I_1 - 3) \tag{6-78}$$

which is functionally identical the Gaussian free energy of deformation [equation (6-72)]. The stress–strain relation can then be obtained from equation (6-73) by differentiation. Again for the special case of uniaxial extension, using equation (6-50) and setting $C_{100} = C_1$, and observing the incompressibility condition, it follows that

$$\sigma - \frac{\partial \overline{W}}{\partial \alpha^*} - 2C_1\left(\alpha^* - \frac{1}{\alpha^{*2}}\right) \tag{6-79}$$

which is known as the neo-Hookean equation. If $C_1 = G_0/2$ equation (6-79) is just the statistical expression [equation (6-60)]. Suppose we retain an additional term with $i = 0$, $j = 1$, $k = 0$, then

$$\overline{W} = C_{100}(I_1 - 3) + C_{010}(I_2 - 3) \tag{6-80}$$

For uniaxial extension, we obtain

$$\sigma = 2C_1\left(\alpha^* - \frac{1}{\alpha^{*2}}\right) + 2C_2\left(1 - \frac{1}{\alpha^{*3}}\right) \tag{6-81}$$

where again for simplicity we have set $C_1 = C_{100}$ and $C_2 = C_{010}$. Equation (6-81) is known as the Mooney–Rivlin equation, which can alternatively be expressed as

$$\sigma = 2\left(C_1 + \frac{C_2}{\lambda}\right)\left(\lambda - \frac{1}{\lambda^2}\right) \tag{6-82}$$

since $\lambda = \alpha^*$ in the incompressible case.

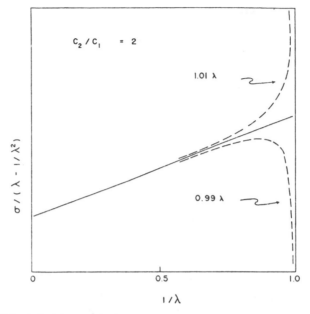

Figure 6-6. Schematic Mooney–Rivlin plot, equation (6-82) for $C_2/C_1=2$. Dotted lines indicate curves with 1% error in λ. Reprinted from B. M. E. Van der Hoff and E. J. Bickler *J. Macromol. Sci.—Chem.*, **A1**, 747 (1967). Courtesy of Marcel Dekker, Inc.

The most sensitive way to test the agreement between equation (6-81) and experimental stress–strain data is to plot the quantity $\sigma/(\lambda-1/\lambda^2)$ against $1/\lambda$. The plot should yield a straight line. Its intercept at $1/\lambda=0$ is $2C_1$ and its slope is $2C_2$. Figure 6-6 shows a schematic Mooney–Rivlin plot[17] for $C_2/C_1=2$. At low strains both $(\lambda-1/\lambda^2)$ and $1/\lambda$ are extremely sensitive to experimental errors in extension ratio measurements. Thus if the measured λ is off from the true λ by only 1%, significant deviations from linearity can be expected in this region of strain. The dashed lines in Figure 6-6 illustrate this behavior.

D.   FACTORS AFFECTING RUBBER ELASTICITY

1.   Effect of Degree of Crosslinking

According to the statistical theory of rubber elasticity, the elastic stress of a rubber under uniaxial extension is directly proportional to the concentration of network chains N_0. For a rubber sample whose density is $d(g/cm^3)$, if the molecular weight of each chain in the network is on the average M_c(g/mole), then $N_0 = d/M_c$ (mole/cm^3). Thus the shear modulus of the rubber [equation (6-59)] can be written as

$$G_0 = \frac{dRT}{M_c} \frac{\overline{r_0^2}}{\overline{r_f^2}} \qquad (6\text{-}83)$$

Equation (6-83) assumes that the network is a perfect one in that all chains in the network are effective in giving rise to the elastic stress. Ideally each crosslink connects four network chains, while each such chain is terminated by two crosslinks. However, as illustrated in Figure 6-2, a number of network imperfections are possible. Each linear polymer chain with molecular weight M, even if all crosslinks are "normal," must give rise to two terminal chains that are incapable of supporting stress. Thus the number of effective chains must not include the imperfections due to chain ends. The shear modulus of the rubber should really be:[18]

$$G_0 = RT \frac{\overline{r_0^2}}{\overline{r_f^2}} \left(\frac{d}{M_c} - \frac{2d}{M} \right) = \frac{dRT}{M_c} \frac{\overline{r_0^2}}{\overline{r_f^2}} \left(1 - \frac{2M_c}{M} \right) \qquad (6\text{-}84)$$

Note, however, that if the initial molecular weight of the linear polymer is infinite, equation (6-84) reduces to equation (6-83). Figure 6-7 shows a plot of G_0 against $1/M$ for a series of natural rubber samples, depicting excellent linearity.[19]

Another type of deviation from the "normal" crosslink structure of the rubber is the effect of chain entanglements (Figure 6-2). Such entanglements would impose additional conformational restrictions on the network chains and thus would have the effect of a "quasi-crosslink" in increasing the elastic stress. Since in actual rubbers the chains are rather closely packed together, one might expect several such entanglements to occur between the crosslinks. Thus their contribution to stress may be quite significant, especially for chains that are sufficiently long to permit a number of such entanglements. In the absense of an effective way to calculate this quantity, we simply add its contribution to the shear modulus as:

$$G_0 = \left(\frac{dRT}{M_c} + a \right) \frac{\overline{r_0^2}}{\overline{r_f^2}} \left(1 - \frac{2M_c}{M} \right) \qquad (6\text{-}85)$$

where a represents the entanglement contribution.

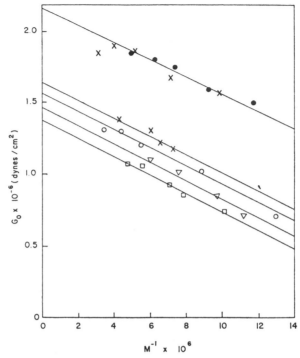

Figure 6-7. Variation of shear moduli of natural rubber with reciprocal initial molecular weight for various vulcanizations. After L. Mullins, *J. Appl. Polym. Sci.*, **2**, 257 (1959), by the permission of John Wiley & Sons, Inc.

In addition to terminal chains and entanglements, there are other types of network imperfections. Figure 6-2 shows that if a short chain were crosslinked only once, the crosslink is a wasted one since the chain cannot support elastic stress. Also, if a crosslink forms an intrachain loop, it is again an ineffective crosslink. Unfortunately, owing to its very complexity, it is at present impossible to completely characterize the network structure of a rubber.

2. Effect of Swelling

Linear polymers are capable of dissolving in appropriate solvents to form homogeneous polymer solutions. However, if crosslinks are introduced to tie the chains into an infinite network, the polymer can no longer dissolve. Instead the solvent is absorbed into the polymer network, giving rise to the phenomenon of swelling. A swollen rubber is in fact a solution, except that its mechanical response is now elastic rather than viscous. As solvents fill the network, chains are extended. The resulting retractive force operates in opposition to the swelling force. There is a maximum degree of swelling, at which point these two forces are at equilibrium.

If the rubber is swollen to below the equilibrium swelling so that no de-swelling will occur upon deformation, the statistical expression for the shear modulus [equation (6-59)] can be readily modified. We define V_r as the ratio of the unswollen volume to the swollen one. The number of network chains per unit volume then becomes $N_0 V_r$, and the mean square end-to-end distance of the network chain is now $\overline{r_0^2}/V_r^{2/3}$. Equation (6-59) then reads

$$G_s = N_0 RTV_r^{1/3} \frac{\overline{r_0^2}}{\overline{r_f^2}} \tag{6-86}$$

for swollen rubbers. The equation of state is now

$$\sigma_s = N_0 RTV_r^{1/3} \left[\frac{\overline{r_0^2}}{\overline{r_f^2}} \right] \left(\lambda_s - \frac{1}{\lambda_s^2} \right) \tag{6-87}$$

where subscripts s refer to the swollen sample. If stress is expressed in terms of per-unit cross-sectional area of *unswollen* sample, then

$$\sigma_d = N_0 RTV_r^{-1/3} \frac{\overline{r_0^2}}{\overline{r_f^2}} \left(\lambda_s - \frac{1}{\lambda_s^2} \right) \tag{6-88}$$

since $A_{0d} = A_{0s} V_r^{2/3}$, where subscripts d refer to the dry (unswollen) sample.

We can also derive an expression on the basis of the Mooney–Rivlin strain energy function for swollen rubbers. A dry rubber sample will undergo two types of deformation: one due to swelling and the other due to extension. The strain energy function per unit volume of swollen rubber is related to that of the dry sample by

$$\overline{W}_s = V_r \overline{W}_d = V_r \left[C_1(I_1 - 3) + C_2(I_2 - 3) \right] \tag{6-89}$$

where subscripts s and d refer to swollen and dry samples respectively. In equation (6-89) the strain invariants I_1 and I_2 are defined by λ_d's, i.e., strains suffered by the dry rubber (both swelling and extension). The deformation due to isotropic swelling is just $V_r^{-1/3}$ for all three principal axes, thus the λ_d's are related to the λ_s's (deformation of swollen rubber by extension) by

$$\lambda_{1d}=\lambda_{1s}V_r^{-1/3} \qquad \lambda_{2d}=\lambda_{2s}V_r^{-1/3} \qquad \lambda_{3d}=\lambda_{3s}V_r^{-1/3} \qquad (6\text{-}90)$$

For the case of simple uniaxial extension, assuming incompressibility for the sake of convenience, we obtain with the aid of equations (6-49) and (6-50)

$$\overline{W}_s = C_1 V_r^{1/3}\left(\lambda_s^2+\frac{2}{\lambda_s}-3\right)+C_2 V_r^{5/3}\left(2\lambda_s+\frac{1}{\lambda_s^2}-3\right) \qquad (6\text{-}91)$$

The stress–strain relation can thus be obtained directly by differentiating with respect to λ_s [equation (6-73)]:

$$\sigma_s = 2C_1 V_r^{1/3}\left(\lambda_s-\frac{1}{\lambda_s^2}\right)+2C_2 V_r^{5/3}\left(1-\frac{1}{\lambda_s^3}\right) \qquad (6\text{-}92)$$

or, in terms of per-unit cross-sectional area of unswollen sample:[20]

$$\sigma_d = 2V_r^{-1/3}\left(C_1+\frac{C_2 V_r^{4/3}}{\lambda_s}\right)\left(\lambda_s-\frac{1}{\lambda_s}\right) \qquad (6\text{-}93)$$

The C_1 term of the Mooney–Rivlin equation is often identified with the shear modulus of the statistical equation; we see that they both depend on V_r in the same manner [compare equation (6-92) with (6-89) or (6-93) with (6-88)].

For comparison with experimental data, we plot $\sigma_d V_r^{1/3}/2(\lambda_s-1/\lambda_s^2)$ as a function of $V_r^{4/3}/\lambda_s$. In Figure 6-8 we see that for natural rubbers swollen to various degrees, data all fall on a straight line with the same slope (C_2) and intercept (C_1) in excellent agreement with theory. The upturns in the region of large strains are due to strain-induced crystallization and finite extensibility of the polymer chains, which are not taken into account by either statistical or phenomenological theories.

3. Effect of Fillers

The use of fillers in rubbers is of paramount technological significance. Such items as automotive tires depend on the addition of fillers to confer on them enhanced wear resistance, strength, and elastic modulus. A wide variety of fillers are commonly employed, such as carbon black, zinc oxide,

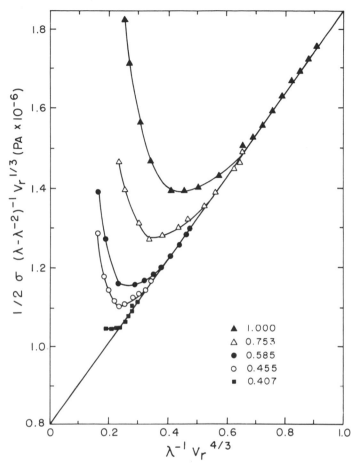

Figure 6-8. Effect of swelling on the Mooney–Rivlin plot of natural rubber where V_r is the volume fraction of rubber in the swollen sample, (swelling liquid, n-decane). After Mullins, *J. Appl. Polym. Sci.*, **2**, 257 (1959), by the permission of John Wiley & Sons, Inc.

carbonates and silicates of calcium and magnesium. Generally there are two types of filler: reinforcing and nonreinforcing. The filler is reinforcing if it is capable of increasing the stiffness of the rubber without impairing its strength and losing its rubbery character.

The most commonly used expression for describing the effect of fillers on the elastic modulus of rubbers is the Guth–Smallwood equation:[22,23]

$$\frac{E_f}{E_0} = 1 + 2.5 V_f + 14.1 V_f^2 \tag{6-94}$$

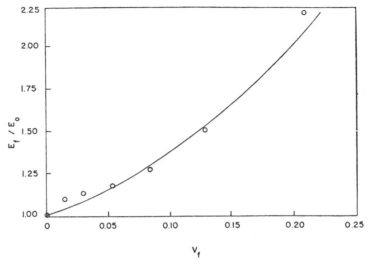

Figure 6-9. Effect of filler on the moduli of natural rubber samples (filler: MT carbon black). The curve was calculated from equation (6-94). After L. Mullins and N. R. Tobin, *J. Appl. Polym. Sci.*, **9**, 2993 (1965), by permission of John Wiley & Sons, Inc.

where V_f is the volume fraction of fillers, and subscripts f and 0 refer to the filled and unfilled rubbers respectively. Equation (6-94) has been found to be valid for a number of filled systems up to a value of V_f of about 0.3. It was actually first derived for the viscosity of liquids with solid suspensions, based on hydrodynamic principles. It was "borrowed" to be used for the elasticity of filled rubbers. The justification for this is to be found in its good agreement with experimental data,[24] as illustrated in Figure 6-9.

The right side of equation (6-94) is sometimes referred to as the strain amplification factor. Suppose we have a sample of rubber in which is dispersed fillers of concentration V_f. The fillers are rigid solid particles with very high modulus. When the filled rubber is deformed, since the fillers are so much more rigid than the rubber, they essentially remain undeformed. Thus all the strain must be suffered by the rubbery phase of the sample. The actual strain sustained by the rubber is greater than the applied strain by the factor given by the right side of equation (6-94). Thus we can write the extension ratio appropriate to the rubber matrix as[24]

$$\Lambda = 1 + \varepsilon \left(1 + 2.5\,V_f + 14.1\,V_f^2\right) \tag{6-95}$$

where ε is the applied strain on the filled sample.

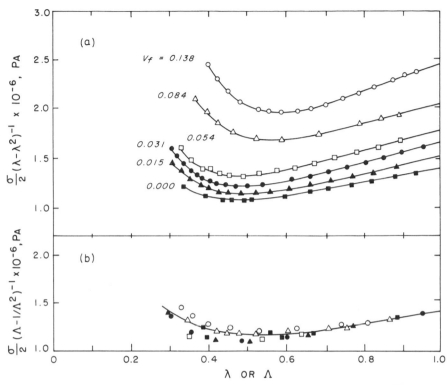

Figure 6-10. Mooney–Rivlin plots of natural rubber filled with MT carbon black: (a) without using the strain amplification factor; (b) using the strain amplification factor, equation (6-95). After L. Mullins and N. R. Tobin, *J. Appl. Polym. Sci.*, **9**, 2993 (1965), by permission of John Wiley & Sons, Inc.

Figure 6-10a shows the Mooney–Rivlin plot of a series of natural rubbers filled with carbon black on the basis of the actually applied strain on the filled sample. As expected, the stress increases as a function of increasing filler concentration. The rise at high strains can again be attributed to crystallization or finite extensibility of the chains in the filled sample. Now if we plot the same data on the basis of the strain sustained by the rubber matrix (Λ), using the strain amplification factor according to equation (6-95), the curves in Figure 6-10b are seen to be brought much closer together. From the slopes and intercepts of these curves, C_1 and C_2 can be readily obtained. The values of these Mooney–Rivlin parameters for the filled samples are quite close to those of the unfilled sample. The agreement is excellent for rubbers filled up to 8%; it is poorer for higher filled contents. These results strongly suggest that the role of fillers is to

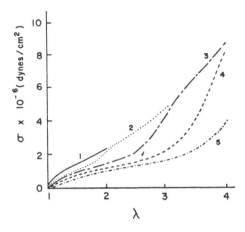

Figure 6-11. Stress-softening of natural rubber filled with MPC carbon black (Mullins effect). Numerals indicate the stress–strain cycles. After F. Bueche, *J. Appl. Polym. Sci.*, **4**, 107 (1960) by permission of John Wiley & Sons.

amplify the applied strain according to equation (6-95). The elastic properties of the rubber itself are unaffected by the presence of these fillers.

Another important effect of fillers is stress-softening, or the Mullins effect. If a filled sample is stretched for the first time to 100%, the stress–strain curve[25] will follow that illustrated in Figure 6-11. Now the strain is removed, and the sample is restretched to 200%. The stress in the second cycle is lower than that in the first up to 100%, after which it continues in a manner following the first cycle. If we repeat the stress–strain in a third cycle, we again see a softening up to 200% due to the previous strain history. This stress-softening effect was first discovered by Mullins, after whom it is named.

A molecular interpretation was first given by F. Bueche[26] for the Mullins effect. Figure 6-12 shows that for a reinforced rubber, the polymer chains are attached to the filler particles. Some of the chains may be relaxed, but others are already relatively extended. Upon stretching, the "prestrained" chains will reach maximum extension first and either will become detached from the fillers or will be broken. In the second cycle, these broken chains no longer support the stress, thus giving rise to the observed softening in stress. If the stress–strain cycle is repeated for the third time, the same process will be repeated. The softened rubber can be "healed" by annealing at higher temperatures.

4.   Effect of Strain-Induced Crystallization

When a sample of rubber is stretched, it becomes anisotropic in that the network chains tend to orient themselves more in the direction of stretch

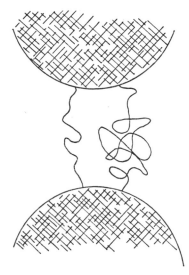

Figure 6-12. Schematic diagram of polymer chains attached to filler particles. After F. Bueche, *Physical Properties of Polymers*, Interscience, New York 1962, p. 49.

than in the lateral directions. Thus more chains are ordered to favor the formation of crystallites. These crystallites will tie together a number of neighboring network chains, thereby exerting a crosslinking effect. This increase in the degree of crosslinking will then in turn cause a rise in the elastic stress. The reason that these crystallites in fact act as crosslinks is attributable to their high modulus, which is estimated to be of the order of 10^{11} Pa. This value is some five orders of magnitude greater than that of most rubbers, which is in the range of 10^6 Pa. If crystallization is allowed to proceed further, more and more of the amorphous material is replaced. At high degrees of crystallinity it is not adequate to regard crystallites just as crosslinks. They will also now act as fillers in further increasing the elastic stress.

That crystallization increases the elastic stress has already been demonstrated in Figure 6-8, in which the Mooney–Rivlin plot shows a rise at high extension ratios. However, it should be remembered that part of this increase is due to finite extensibility of network chains. In Figure 6-13 we show the stress–strain curves of natural rubber at two temperatures.[27] At 0°C there is considerable strain-induced crystallization, and we observe a dramatic rise in the elastic stress above $\lambda = 3.0$. X-ray measurements show the appearance of crystallinity above this strain. At 60°C there is little or no crystallization, and the stress–strain curve shows a much smaller upturn at high strains. The latter is presumably due only to the finite extensibility of the polymer chains in the network.

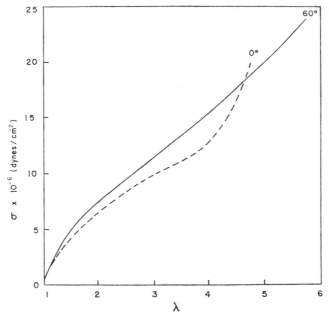

Figure 6-13. Stress–strain curves of natural rubber at 0°C and 60°C. The rise in stress at 0°C at $\lambda > 3$ is attributed to strain-induced crystallization. After K. J. Smith, A. Greene, and A. Ciferri, *Kolloid-Z*, **194**, 49 (1964), by permission of Dr. Dietrich Steinkopff Verlag.

PROBLEMS

1. Starting with the equation of state of rubber elasticity [equation (6-60)], show that for an ideal rubber

$$\text{(a)} \quad \left(\frac{\partial H}{\partial L}\right)_{T,P} = \frac{\alpha T}{3}\left(\frac{\lambda^3 + 2}{\lambda^3 - 1}\right)$$

$$\text{(b)} \quad \left(\frac{\partial U}{\partial L}\right)_{T,V} = 0$$

$$\text{(c)} \quad \left(\frac{\partial V}{\partial L}\right)_{T,P} = \frac{\beta}{3}\left(\frac{\lambda^3 + 2}{\lambda^3 - 1}\right)$$

2. From the above three equations and equation (6-60), find an expression for the stress–temperature coefficient of an ideal rubber at constant pressure and length, that is $(\partial f/\partial T)_{P,L}$.

3. If a rubber sample in the form of a unit cube is deformed by pure shear, then the three principal extension ratios are $\lambda_1 = \lambda$, $\lambda_2 = 1$, $\lambda_3 = 1/\lambda$. [Compare with the case of simple extension in equation (6-50).] Following the arguments of Section B, derive an expression relating σ and λ.

4. Repeat Problem 3 using the Mooney–Rivlin strain energy function [equation (6-80)].

5. In analogy to the kinetic theory of ideal gases, the statistical theory of rubber elasticity is often called the kinetic theory of rubber elasticity. Reflect upon the similarities and differences between the basic philosophies of these two theories.

6. For a piece of ideal rubber, whose density is 0.95 g/cm^3 calculate its shear modulus at room temperature if its initial molecular weight is 100,000 g/mole and the molecular weight of the network chain after crosslinking is 5000 g/mole (assuming absence of other network defects).

7. What is the maximum degree of swelling for a rubber that would fail at 100% extension in the unswollen state?

8. If the tensile modulus for an unfilled rubber is 5×10^7 dynes/cm^2, what value would it have if it is filled to $V_f = 0.3$? Suppose after several strain cycles, half of its fillers are ineffective due to Mullins effect. What is its tensile modulus now?

9. Starting with equation (6-66), derive equation (6-67).

REFERENCES

1. M. Shen, D. A. McQuarrie, and J. L. Jackson, *J. Appl. Phys.*, **38**, 791 (1967).
2. G. Gee, J. Stern, and L. R. G. Treloar, *Trans. Faraday Soc.*, **46**, 1101 (1950).
3. F. G. Hewitt and R. L. Anthony, *J. Appl. Phys.*, **29**, 1411 (1958).
4. G. Allen, U. Bianchi, and C. Price, *Trans Faraday Soc.*, **59**, Part II, 2493 (1963).
5. E. Guth and H. F. Mark, *Monatsh.*, **65**, 93 (1934).
6. W. Kuhn, *Kolloid-Z*, **68**, 2 (1934); *Kolloid-Z.*, **76**, 258 (1936).
7. F. T. Wall, *J. Chem. Phys.*, **10**, 132, 485 (1942); *J. Chem. Phys.*, **11**, 527 (1943).
8. H. M. James and E. Guth, *J. Chem. Phys.*, **11**, 455 (1943); *J. Polym. Sci.*, **4**, 153 (1949).
9. P. J. Flory and J. Rehner, Jr., *J. Chem. Phys.*, **11**, 512, 521 (1943); P. J. Flory, *J. Am. Chem. Soc.*, **18**, 5232 (1956); *Trans. Faraday Soc.*, **56**, 722 (1960).
10. M. S. Green and A. V. Tobolsky, *J. Chem. Phys.*, **14**, 80 (1946); A. V. Tobolsky, D. W. Carlson, and N. Indictor, *J. Polym. Sci.*, **54**, 175 (1961).
11. J. J. Hermans, *Trans. Faraday Soc.*, **43**, 591 (1947); *J. Polym. Sci.*, **59**, 191 (1962).
12. P. J. Flory, *Proc. R. Soc. Lond.*, **A351**, 351 (1976).
13. L. R. G. Treloar, *Trans. Faraday Soc.*, **40**, 59 (1944).
14. M. Shen and P. J. Blatz, *J. Appl. Phys.*, **39**, 4937 (1968); M. Shen, *Macromolecules*, **2**, 358 (1969).
15. M. Mooney, *J. Appl. Phys.*, **11**, 582 (1940).
16. R. S. Rivlin, *Philos. Trans.*, **A241**, 379 (1948).
17. B. M. E. Van der Hoff and E. J. Buckler, *J. Macromol. Sci.–Chem.*, **A1**, 747 (1967).
18. P. J. Flory, *Chem. Revs.*, **35**, 51 (1944).

19. L. Mullins, *J. Appl. Polym. Sci.*, **2**, 1 (1959).
20. B. M. E. Van der Hoff, *Polym. Sci.*, **6**, 397 (1965).
21. L. Mullins, *J. Appl. Polym. Sci.*, **2**, 257 (1959).
22. E. Guth, R. Simha, and O. Gold, *Kolloid-Z.*, **74**, 266 (1936); E. Guth, *J. Appl. Phys.*, **16**, 20 (1945).
23. H. M. Smallwood, *J. Appl. Phys.*, **15**, 758 (1944).
24. L. Mullins and N. R. Tobin, *J. Appl. Polym. Sci.*, **9**, 2993 (1965).
25. L. Mullins and N. R. Tobin, *Trans. IRI*, **33**, 2 (1956).
26. F. Bueche, *J. Appl. Polym. Sci.*, **4**, 107 (1960); *J. Appl. Polym. Sci.*, **5**, 27 (1961).
27. K. J. Smith, A. Greene, and A. Ciferri, *Kolloid-Z.*, **194**, 49 (1964).

7

Viscoelastic Models

The phenomenological theory of linear viscoelasticity developed in Chapter 2 is completely independent of the existence of models. It is desirable to consider the representation of linear viscoelastic processes by certain model systems in order to gain greater insight into relaxation behavior. In this chapter we consider two broad classes of models. The first consists of the so-called "mechanical analogues." These are combinations of elements, usually springs and dashpots, that more or less faithfully reproduce the viscoelastic response of real systems. The second group is composed of the molecular theories. Here a fairly reasonable representation of a polymer molecule is assumed and the motion of such a molecule in a viscous medium is deduced. In this case the viscoelastic behavior is predicted on the basis of molecular parameters. It will be demonstrated that the two classes of models are equivalent in many respects.

A. MECHANICAL ELEMENTS

We return to the tensile elongation experiment described in the Introduction and Chapter 2. The simplest mechanical model that has some of the gross physical behavior exhibited by bodies subject to uniaxial elongation is a pure Hookean spring (Figure 7-1a). This body is purely elastic and all inertial effects are neglected. Thus if the Hookean spring is subjected to an instantaneous stress σ_0, it will respond instantaneously with a strain ε_0, σ_0 and ε_0 being related by the equation

$$\sigma_0 = E\varepsilon_0 \tag{7-1}$$

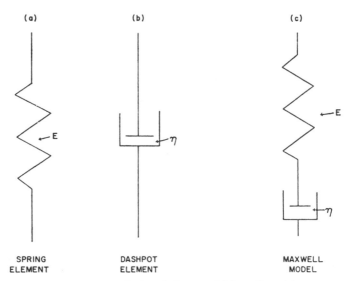

Figure 7-1. Spring, dashpot, and Maxwell model.

The proportionality constant, E, is the Young's modulus with which we are already acquainted. The instantaneous stress application is assumed to produce no oscillations, only a constant strain. It is obvious that no actual substances obey Hooke's law. Some materials, such as steel, obey Hooke's law over a wide range of stress and strain, as stated in equation (7-1), but none respond without inertial effects.

The dominant characteristic of fluids, on the other hand, is not their elasticity, but rather their viscosity. Newton's law

$$\sigma = \eta \frac{d\varepsilon}{dt} \tag{7-2}$$

is the equation of motion for a model with a simple linear viscous behavior. The mechanical analogue of equation (7-2) is the dashpot element (Figure 7-1b). This is merely a piston in a cyclinder filled with a liquid of viscosity η. Integration of equation (7-2) for constant stress σ_0, yields

$$\varepsilon(t) = \frac{\sigma_0}{\eta} t \tag{7-3}$$

Thus subjecting the dashpot to a stress of $10\sigma_0$ for time t produces the same strain as a stress of σ_0 applied for a time $10t$.

1. Maxwell Model

The mechanical response of viscoelastic bodies such as polymers are poorly represented by either the spring or the dashpot. J. C. Maxwell suggested that a better approximation would result from a series combination of the spring and dashpot elements. Such a model, called a Maxwell element, is shown in Figure 7-1c. In the Maxwell element, E, the instantaneous tensile modulus, characterizes the response of the spring while η, the viscosity of the liquid in the dashpot, defines the viscous behavior. One can stipulate a relationship between E and η, without any loss of generality,

$$\eta = \tau E \tag{7-4}$$

τ, the proportionality constant between E and η, is known as the relaxation time of the element. The equation of motion of the Maxwell model is

$$\frac{d\varepsilon}{dt} = \frac{1}{E}\frac{d\sigma}{dt} + \frac{\sigma}{\eta} \tag{7-5}$$

Equation (7-5) is merely a linear combination of perfectly elastic behavior as stated in the time derivative of Hooke's law (first term on right side) and perfectly viscous behavior as stated in Newton's law (second term on right side).

We will now solve this differential equation subject to several sets of experimental boundary conditions.

Creep Experiment. The model is subjected to an instantaneous constant stress σ_0. Thus equation (7-5) becomes

$$\frac{d\varepsilon}{dt} = \frac{\sigma_0}{\eta} \tag{7-6}$$

since $d\sigma/dt$ is zero. Integration of equation (7-6) from time 0 to some time t yields, after division by σ_0

$$\frac{\varepsilon(t)}{\sigma_0} = \frac{\varepsilon_0}{\sigma_0} + \frac{t}{\eta} \tag{7-7}$$

One notices that ε_0/σ_0 is just the instantaneous response of the Hookean spring, and from equation (7-1) it is seen to be equal to the reciprocal of E.

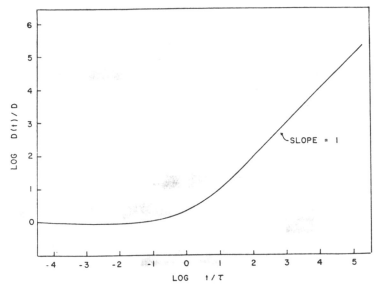

Figure 7-2. Maxwell body behavior using creep conditions; log–log plot.

From the results of Chapter 2, Section A, we have

$$D = \frac{1}{E} = \frac{\varepsilon_0}{\sigma_0} \tag{7-8}$$

where D is the tensile compliance of the spring.

Also from the results of Chapter 2, Section A, it is clear that the left side of equation (7-7) is simply the tensile creep compliance, $D(t)$. Thus the response of the Maxwell model to a creep experiment is

$$D(t) = D + \frac{t}{\eta} \tag{7-9}$$

This is illustrated in Figure 7-2.

Stress Relaxation Experiment. In a stress relaxation experiment, one strains the sample instantaneously to some strain ε_0 and studies the stress $\sigma(t)$ necessary to maintain this constant strain (Chapter 2, Section B). The instantaneous strain will be realized only in the spring element. The dashpot will gradually relax so that the stress decreases as a function of time.

After the strain application, $(d\varepsilon/dt)$ is zero so equation (7-5) becomes, in light of equation (7-4),

$$\frac{d\sigma}{\sigma} = d\ln\sigma = \frac{-dt}{\tau} \tag{7-10}$$

This expression is easily integrated from σ_0 at time 0 to $\sigma(t)$ at time t to give

$$\ln\sigma(t) = \ln\sigma_0 - \frac{t}{\tau} \tag{7-11}$$

Exponentiation and division by ε_0 yields

$$\frac{\sigma(t)}{\varepsilon_0} = \frac{\sigma_0}{\varepsilon_0}e^{-t/\tau} \tag{7-12}$$

Here again, σ_0/ε_0 is the modulus of the spring, E, and the left side of equation (7-12) is $E(t)$, the tensile stress relaxation modulus.

$$E(t) = Ee^{-t/\tau} \tag{7-13}$$

Figures 7-3a and 7-3b illustrate the behavior of the Maxwell model in a stress relation experiment. Note that these functions have been plotted on log–log plots as well as the usual rectilinear scales. At times considerably shorter than the relaxation time of the dashpot, the element behaves as if it were a spring alone. At times very long compared to the relaxation time of the dashpot, the model behaves as if it were a dashpot alone, that is, the stress decays to zero. At times comparable to the relaxation time, the response involves both the spring and the dashpot.

Dynamic Experiments. We will now consider the response of a Maxwell element subjected to a sinusoidal stress. In such a case the strain will also be sinusoidal but out of phase with the stress by the angle δ, as discussed in Chapter 2. Thus

$$\sigma(t) = \sigma_0 e^{i\omega t} \tag{7-14}$$

where σ_0 is the amplitude of the stress and ω is the frequency. Substitution into equation (7-5) yields

$$\frac{d\varepsilon(t)}{dt} = \frac{\sigma_0}{E}i\omega e^{i\omega t} + \frac{\sigma_0}{\eta}e^{i\omega t} \tag{7-15}$$

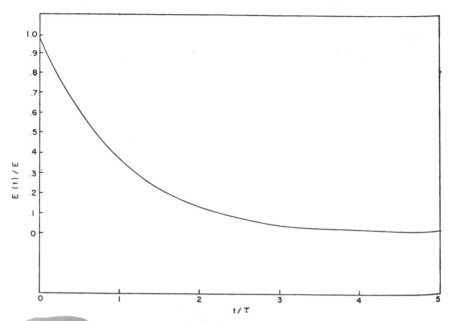

Figure 7-3a. Maxwell body behavior using stress relaxation conditions; linear plot.

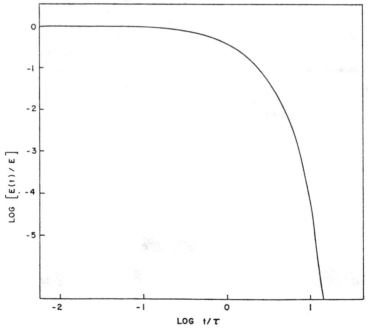

Figure 7-3b. Maxwell body behavior using stress relaxation conditions; log–log plot.

144

which is easily integrated using the limits of $\varepsilon(t_1)$ at t_1 and $\varepsilon(t_2)$ at t_2. [Note that $\varepsilon(0)$ is not necessarily zero since the stress and strain are not in phase.]

$$\varepsilon(t_2) - \varepsilon(t_1) = \frac{\sigma_0}{E}(e^{i\omega t_2} - e^{i\omega t_1}) + \frac{\sigma_0}{\eta i\omega}(e^{i\omega t_2} - e^{i\omega t_1}) \qquad (7\text{-}16)$$

Division of the strain increment by the stress increment yields:

$$\frac{\varepsilon(t_2) - \varepsilon(t_1)}{\sigma(t_2) - \sigma(t_1)} = D^* = D - i\frac{D}{\tau\omega} \qquad (7\text{-}17)$$

where D^* is the complex tensile creep compliance. In terms of storage and loss compliances (Chapter 2),

$$D^* = D' - iD''$$

$$\qquad (7\text{-}18)$$

$$D' = D \quad \text{and} \quad D'' = \frac{D}{\tau\omega} = \frac{1}{\eta\omega}$$

From this point, the calculation of the complex tensile modulus, E^*, is straightforward. From Chapter 2, Section C, it will be remembered that E^* is the reciprocal of D^*, so we have

$$E^* = \frac{1}{D - iD/\tau\omega} = \frac{\tau\omega E}{\tau\omega - i} \qquad (7\text{-}19)$$

Use of the complex conjugate gives

$$E^* = \frac{E\tau^2\omega^2}{1 + \omega^2\tau^2} + \frac{i\tau\omega E}{1 + \omega^2\tau^2} = E' + iE'' \qquad (7\text{-}20)$$

The tangent of the phase angle between the stress and strain was shown to be (Chapter 2, Section C):

$$\tan\delta = \frac{E''}{E'} \qquad (2\text{-}21)$$

so

$$\tan\delta = \frac{1}{\tau\omega}$$

$$E' = \frac{E\tau^2\omega^2}{1 + \omega^2\tau^2} \qquad (7\text{-}21)$$

$$E'' = \frac{E\tau\omega}{1 + \omega^2\tau^2}$$

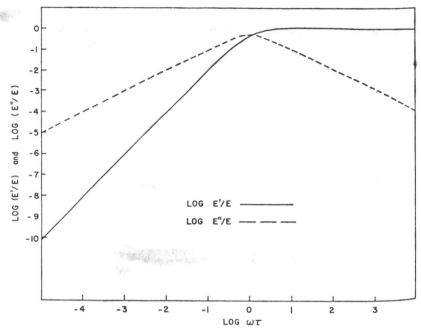

Figure 7-4.   Log–log plots of $E'(\omega)/E$ and $E''(\omega)/E$ versus $\omega\tau$ for a Maxwell model.

for a Maxwell body. Figure 7-4 represents the frequency dependence of E' and E'' for a Maxwell element.

2. Voigt Element

Another simple element has been used frequently in connection with viscoelastic behavior. The so-called Voigt model consists of the same fundamental elements as the Maxwell model, except here the spring and dashpot are in parallel instead of being in series (Figure 7-5). The constraint on the model is that the strain must be the same in both elements. The stress then must be the sum of the stresses in the two individual elements. Thus the fundamental equation of motion for the Voigt element is

$$\sigma(t) = \varepsilon(t)E + \eta\frac{d\varepsilon(t)}{dt} \tag{7-22}$$

This model is usually used in considering creep experiments since, as will be seen shortly, it cannot easily be applied to a stress relaxation experiment.

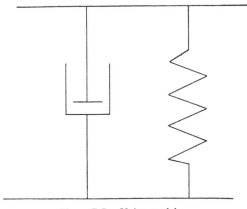

Figure 7-5.   Voigt model.

Creep Experiment.   Since the stress is a constant in a creep experiment, one has

$$\frac{d\varepsilon(t)}{dt} + \frac{\varepsilon(t)}{\tau} = \frac{\sigma_0}{\eta} \tag{7-23}$$

Equation (7-23) is a linear differential equation which can be made exact and then integrated using the integrating factor $e^{t/\tau}$. Integration between the limits $\varepsilon(0) = 0$ and $\varepsilon(t) = \varepsilon(t)$ yields

$$\int_{\varepsilon(0)}^{\varepsilon(t)} d\left[\varepsilon(t)e^{t/\tau}\right] = \frac{\sigma_0}{\eta}\int_0^t e^{t/\tau}\, dt$$

$$\tag{7-24}$$

$$\varepsilon(t)e^{t/\tau} = \frac{\sigma_0}{E}(e^{t/\tau} - 1)$$

or

$$\frac{\varepsilon(t)}{\sigma_0} = D(t) = D(1 - e^{-t/\tau})$$

Stress Relaxation.   With constant strain, the equation of motion of the Voigt element reduces to Hooke's law with

$$E(t) = E \tag{7-25}$$

It is clear, however, that true stress relaxation is an impossible experiment

**Table 7.1.   Behaviors of Simple Viscoelastic Models in
Various Experiments**

Experiment	Maxwell Element	Voigt Element
Creep	$D(t) = D + t/\eta$	$D(t) = D(1 - e^{-t/\tau})$
Stress relaxation	$E(t) = Ee^{-t/\tau}$	$D(t) = E$
Sinusoidal dynamic experiments	$D' = D$	$D' = \dfrac{D}{1 + \omega^2\tau^2}$
	$D'' = 1/\eta\omega$	$D'' = \dfrac{D\omega\tau}{1 + \omega^2\tau^2}$
	$E' = \dfrac{E\omega^2\tau^2}{1 + \omega^2\tau^2}$	$E' = E$
	$E'' = \dfrac{E\omega\tau}{1 + \omega^2\tau^2}$	$E'' = \omega\eta$

to carry out on a Voigt element since it would take an infinite force to strain the viscous element instantaneously.

The application of sinusoidal stress and strain is similar to that for a Maxwell body. The results are summarized in Table 7-1 along with the previously derived results for a Maxwell element. Figure 7-6 represents the frequency dependence of D' and D'' for the Voigt element.

The response of both the Maxwell and Voigt models to several kinds of deformation experiments are much simpler than those of real polymer systems according to the results presented in Chapter 3. In particular, whereas most linear polymers of sufficiently high molecular weight have at least two major transitions (glass to rubber and rubber to liquid), these models exhibit only one transition under all conditions. Upon closer examination, it is also apparent that the decay of modulus exhibited by a Maxwell model at times slightly greater than τ is much more rapid than the corresponding modulus decay exhibited by real polymers in either transition region. Thus these simple models do not provide good approximations of the viscoelastic behavior of polymers.

In order to overcome these deficiencies, models have been proposed that consist of combinations of Maxwell and Voigt elements. Although an infinite number of such combinations is possible, we will consider only two here. The treatments of other such models are completely analogous.

3.   Maxwell–Wiechert Model

This generalized model consists of an arbitrary number of Maxwell elements connected in parallel as shown in Figure 7-7.

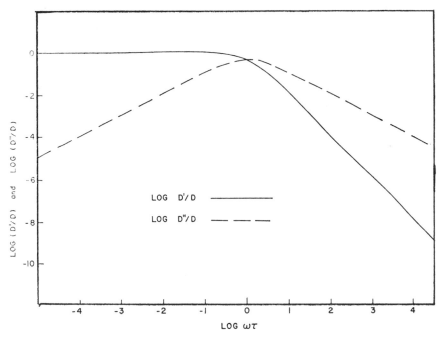

Figure 7-6. Log–log plots of $D'(\omega)/D$ and $D''(\omega)/D$ versus $\omega\tau$ for the Voigt model.

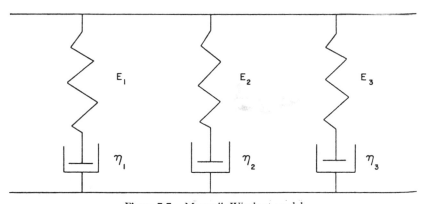

Figure 7-7. Maxwell–Wiechert model.

Consider a Maxwell–Wiechert model with z elements subjected to a stress relaxation experiment. The strain in each element is characterized by a spring constant E_i and a viscosity η_i, thus each τ_i is determined. In all of the individual elements, the strain is the same and the total stress σ is the summation of the individual stresses experienced by each element. One can then write:

$$\frac{d\varepsilon(t)}{dt} = 0 = \frac{1}{E_1}\frac{d\sigma_1}{dt} + \frac{\sigma_1}{\eta_1}$$

$$= \frac{1}{E_2}\frac{d\sigma_2}{dt} + \frac{\sigma_2}{\eta_2}$$

$$= \frac{1}{E_n}\frac{d\sigma_n}{dt} + \frac{\sigma_n}{\eta_n}$$

$$= \frac{1}{E_z}\frac{d\sigma_z}{dt} + \frac{\sigma_z}{\eta_z} \qquad (7\text{-}26)$$

$$(7\text{-}5)$$

$$\sigma = \sigma_1 + \sigma_2 + \sigma_3 + \cdots + \sigma_n + \cdots + \sigma_z \qquad (7\text{-}27)$$

Integration of equation (7-26) gives the partial stresses σ_i, which can then be substituted into equation (7-27) to calculate the total stress. When the total stress is divided by the constant strain, ε_0, the stress relaxation modulus results:

$$E(t) = \frac{\sigma(t)}{\varepsilon_0} = \frac{\sigma(0)_1}{\varepsilon_0}e^{-t/\tau_1} + \frac{\sigma(0)_2}{\varepsilon_0}e^{-t/\tau_2}$$

$$+ \cdots + \frac{\sigma(0)_n}{\varepsilon_0}e^{-t/\tau_n} + \cdots + \frac{\sigma(0)_z}{\varepsilon_0}e^{-t/\tau_z}$$

$$= \sum_{i=1}^{z} E_i e^{-t/\tau_i} \qquad (7\text{-}28)$$

where $\sigma(0)_n$ is the stress on the nth element at time equals zero.

Thus we see that the total modulus is the summation of the responses of the individual elements.

The flexibility of this model in reproducing real viscoelastic behavior can be easily demonstrated. Consider a two-element model with parameters

$$E_1 = 3 \times 10^9 \text{ Pa}; \quad \tau_1 = 1 \text{ min} \quad \text{and} \quad E_2 = 5 \times 10^5 \text{ Pa}; \quad \tau_2 = 10^3 \text{ min}$$

Its behavior (Figure 7-8) reproduces the two transitions observed in real polymers. It is possible to replace one of the Maxwell elements in the Maxwell–Wiechert model with a spring. The stress would decay to a finite

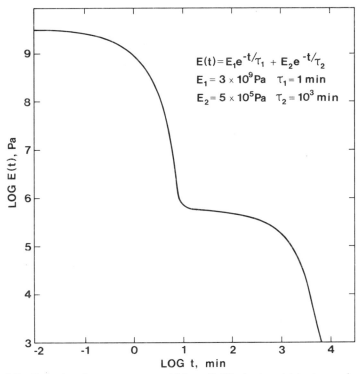

$$E(t) = E_1 e^{-t/\tau_1} + E_2 e^{-t/\tau_2}$$
$$E_1 = 3 \times 10^9 \, Pa \quad \tau_1 = 1 \, min$$
$$E_2 = 5 \times 10^5 \, Pa \quad \tau_2 = 10^3 \, min$$

Figure 7-8. Behavior of a two component Maxwell–Wiechert model in stress relaxation.

value in such a model rather than 0 and would approximate the behavior of crosslinked polymers.

Treatments similar to those used in equations (7-26) and (7-27) can be applied to the Maxwell–Wiechert model undergoing sinusoidal stress or strain to calculate the complex modulus. The results are given in Table 7-2. Expressions for the complex compliance are not simple functions.

4. Voigt–Kelvin Model

The Voigt–Kelvin model is a generalization of the Voigt element that results from connecting Voigt elements in series (Figure 7-9). Here the compliance functions are easily calculated, while the modulus functions are rather complicated. The results are summarized in Table 7-2; a sample calculation is provided below.

Here we derive expressions for D' and D'' of a Voigt–Kelvin model consisting of z elements assuming a sinusoidal strain application.

Table 7-2. Behaviors of Maxwell–Wiechert and Voigt–Kelvin Models in Various Experiments

Experiment	Maxwell–Wiechert Model	Voigt–Kelvin Model
Creep		$D(t) = \sum\limits_{i=1}^{z} D_i(1 - e^{-t/\tau_i})$
Stress relaxation	$E(t) = \sum\limits_{i=1}^{z} E_i e^{-t/\tau_i}$	
Sinusoidal dynamic experiments	$E' = \sum\limits_{i=1}^{z} \dfrac{E_i \omega^2 \tau_i^2}{1 + \omega^2 \tau_i^2}$	$D' = \sum\limits_{i=1}^{z} \dfrac{D_i}{1 + \omega^2 \tau_i^2}$
	$E'' = \sum\limits_{i=1}^{z} \dfrac{E_i \omega \tau_i}{1 + \omega^2 \tau_i^2}$	$D'' = \sum\limits_{i=1}^{z} \dfrac{D_i \omega \tau_i}{1 + \omega^2 \tau_i^2}$

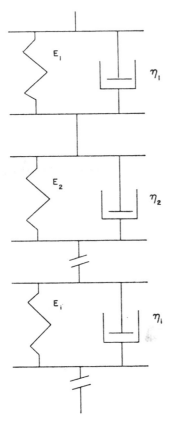

Figure 7-9. Voigt–Kelvin model.

Applying equation (7-22) to the Voigt–Kelvin model experiencing a strain in the jth element given by

$$\varepsilon_j(t) = \varepsilon_{0_j} e^{i\omega t}$$

results in the set of equations

$$\sigma(t) = E_j \varepsilon_{0_j} e^{i\omega t} + \eta_j i \omega \varepsilon_{0_j} e^{i\omega t} \qquad \text{for} \quad j = 1 \text{ to } z \qquad (7\text{-}29)$$

$$\varepsilon(t) = \sum_{j=1}^{z} \varepsilon_j(t) \qquad (7\text{-}30)$$

Solving equations (7-29) and (7-30) for the strain in the jth element gives

$$\varepsilon_j(t) = \varepsilon_{0_j} e^{i\omega t} = \frac{\sigma(t)}{E_j + \eta_j i\omega} \qquad (7\text{-}31)$$

which, when substituted into equation (7-30) and simplified, results in

$$\frac{\varepsilon(t)}{\sigma(t)} = D^* = \sum_{j=1}^{z} \frac{1}{E_j + \eta_j i\omega} = \sum_{j=1}^{z} \frac{D_j}{1 + \omega^2 \tau_j^2} - i \sum_{j=1}^{z} \frac{D_j \omega \tau_j}{1 + \omega^2 \tau_j^2} \qquad (7\text{-}32)$$

It is clear that an inversion of equation (7-32) will not yield a simple result.

B.   DISTRIBUTIONS OF RELAXATION AND RETARDATION TIMES

The pertinent parameters used in a Maxwell–Wiechert model can be easily presented in graphical form as shown in Figure 7-10. Here three spring constants (10^{10}, 10^8, and 10^6) are associated with relaxation times (10^2, 10^4, and 10^6 respectively). Clearly the stress relaxation modulus for this particular Maxwell–Wiechert model could be easily calculated using equation (7-28).

In considering systems where there are very many Maxwell elements employed in the model, that is, z in equation (7-28) is large, it is often convenient to replace the summation in the equation by an integration. Thus:

$$E(t) = \int_0^\infty E(\tau) e^{-t/\tau} \, d\tau \qquad (7\text{-}33)$$

The various E_i's are replaced by a continuous function, $E(\tau)$, of the

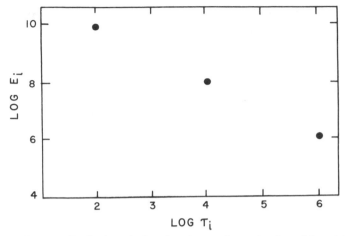

Figure 7-10. Discrete distribution of relaxation times and associated partial modulus values.

relaxation time where this function is called a distribution of relaxation times. In addition to the distribution $E(\tau)$, one often encounters $H(\tau)$, which is simply defined as

$$H(\tau) = \tau E(\tau) \qquad (7\text{-}34)$$

Thus in terms of $H(\tau)$, equation (7-33) becomes:

$$E(t) = \int_0^\infty \frac{H(\tau)}{\tau} e^{-t/\tau} d\tau = \int_{\ln \tau = -\infty}^{\ln \tau = +\infty} H(\tau) e^{-t/\tau} d\ln \tau \qquad (7\text{-}35)$$

Consider calculating the modulus for the distribution of relaxation times given in Figure 7-11. Mathematically $H(\tau)$ can be written as

$$
\begin{aligned}
H(\tau) &= 0 & \log \tau < 0 \\
H(\tau) &= k\tau^{-1} & 0 < \log \tau < 1 \\
H(\tau) &= 0 & \log \tau > 1
\end{aligned}
\qquad (7\text{-}36)
$$

With this $H(\tau)$, equation (7-35) becomes:

$$E(t) = \frac{k}{t}(e^{-(t/10)} - e^{-t}) \qquad (7\text{-}37)$$

This function is also plotted in Figure 7-11.

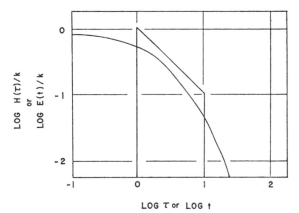

LOG τ or LOG t

Figure 7-11. Continuous distribution of relaxation times expressed as $H(\tau)$ and the corresponding tensile stress relaxation modulus $E(t)$.

In addition, dynamic modulus functions can be calculated via this distribution function; $E'(\omega)$, for example, is given as

$$E'(\omega) = \int_{\ln \tau = -\infty}^{\ln \tau = \infty} H(\tau) \frac{\omega^2 \tau^2}{1 + \omega^2 \tau^2} d\ln \tau \qquad (7\text{-}38)$$

A similar method is used to consider compliance functions. Here, however, the distribution of retardation times $L(\tau)$ defined as

$$J(t) = \int_{\ln \tau = -\infty}^{\ln \tau = \infty} L(\tau)(1 - e^{-t/\tau}) d\ln \tau \qquad (7\text{-}39)$$

is used. Although $H(\tau)$ and $L(\tau)$ are related, their exact quantitative relationship is quite complicated; the interested reader should consult sources such as Ferry[6] or Gross.[18] Mathematical methods for extracting distribution functions from experimental modulus or compliance data are also given by these authors. In addition, see Problem 10 at the end of this chapter.

Distributions of relaxation times for real polymer systems are slightly more complex than that shown in Figure 7-11. Tobolsky has suggested, for example, that the stress relaxation modulus of NBS polyisobutylene shown in Figure 7-12 can be thought to arise from a distribution of relaxation times that is composed of a "box and a wedge." This composite is shown in

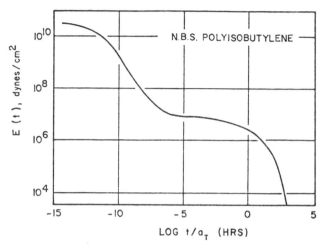

Figure 7-12. Stress relaxation master curve for N.B.S. polyisobutylene at 25°C. After A. V. Tobolsky, *Properties and Structure of Polymers*, p. 151, by permission of John Wiley & Sons, Inc.

Figure 7-13, and can be expressed as

$$H(\tau) = M/\tau^{1/2} \qquad \tau_1 < \tau < \tau_2$$
$$H(\tau) = E_0 \qquad \tau_3 < \tau < \tau_m \qquad (7\text{-}40)$$

In this particular case, Tobolsky gives the pertinent parameters the following values:

$$M = 8.9 \times 10^3 \qquad\qquad E_0 = 7.2 \times 10^5$$

$$\tau_1 = 10^{-12.5} \qquad\qquad \tau_2 = 10^{-5.4}$$

$$\tau_3 = 9.65 \times 10^{-26} \overline{M}_w^{3.30} \qquad\qquad (7\text{-}41)$$

$$\tau_m = 1.06 \times 10^{-20} \overline{M}_w^{3.30}$$

Clearly the wedge is independent of molecular weight and gives rise to the primary transition. The box portion of the spectrum generates the rubbery plateau and rubbery flow regions of the master curve. As discussed in Chapter 3, Section A, these regions are strong functions of the molecular weight, and the box portion of the spectrum mirrors this fact in the dependence of τ_3 and τ_m on molecular weight.

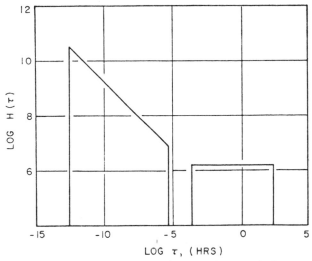

Figure 7-13. Box-and-wedge distribution of relaxation times equivalent to master curve shown in Figure 7-12. After A. V. Tobolsky, *Properties and Structure of Polymers*, p. 128, by permission of John Wiley & Sons, Inc.

C. MOLECULAR THEORIES

Having discussed the viscoelastic responses of simple mechanical models, we may now consider molecular theories. In this treatment it will be shown that the results of molecular theories, can, in fact, be couched in terms of the mechanical models already presented. The molecular theories predict the distribution of relaxation times and partial moduli associated with each relaxation time (τ_i's and E_i's for all i's), which we treated as unknowns or parameters in the previous discussion. Thus, although molecular theories are not based on mechanical models, the results of these treatments may be presented in terms of the parameters of these models. Since, as we have already shown, it is possible to develop expressions giving the viscoelastic responses of the models to various types of deformations, the predictions of the molecular theories are obtainable through the responses of the models.

Although far from being rigorous, a presentation of some of the salient points of the Rouse theory[1] will be attempted using the method of Peticolas.[2] The conclusions reached in this section are more or less applicable to the theories presented by Rouse,[1] Bueche,[3] and Zimm[4] among others. The aim of this treatment is to present assumptions upon which the theory is based and to outline the results. Mathematical rigor is not attempted in this

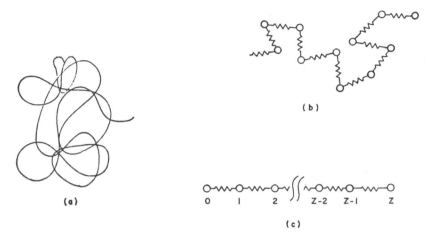

Figure 7-14. Bead-and-spring representation of a real polymer molecule in dilute solution.

introductory treatment. For a complete or perhaps accurate understanding of the field, the reader *must* consult the original papers.[1,3,4]

Having considered the treatment of the isolated polymer chain in Chapter 5, Section E, one realizes that a long freely orienting molecule behaves like a Hookean entropy spring. (Figure 7-14a.) Equation (5-56) shows that the restoring force on this spring when it is distorted by some amount ΔX is given by

$$f = \frac{3kT}{\overline{r^2}}\Delta X \qquad (5\text{-}56)$$

The molecular theory subdivides the polymer molecule into subunits or submolecules just long enough so that the end-to-end distribution of each of these subunits is Gaussian. This is done so that equation (5-56) is applicable to each submolecule. Our picture of the molecule then becomes much like Figure 7-14b. The mass of the submolecule is concentrated at the beads, which are held together by the Hookean springs. Since we will be considering the response of the system to a unidirectional perturbation in the x direction, a spring oriented exactly perpendicular to the x direction will not contribute to the response. Figure 7-15 indicates the deformations allowed for springs in different configurations relative to the perturbing function. It should be clear that an "effective spring constant" in the x direction will now allow us to portray our system as a one-dimensional chain (Figure 7-14c). This linearization of the problem is in no way restrictive. It arises because the polymer chain is linear and because the

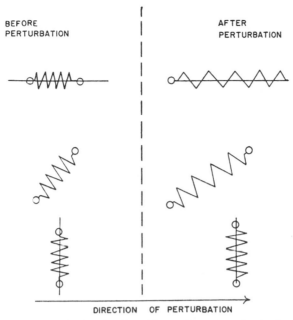

Figure 7-15. Deformations of submolecules with varying orientation to the unidirectional perturbation.

deformation considered is a unidirectional one. The only interactions in the model are those of adjacent submolecules with one another.

Having proceeded this far, we are now in a position to write the equation of motion for our linear bead-and-spring model. In order to do this we divide the polymer molecule into z submolecules so that there are z springs and $z+1$ beads. We now introduce a unidirectional deformation. The restoring force on each of the beads is given by

$$f_{0x} = \frac{-3kT}{a^2}(X_0 - X_1)$$

$$f_{1x} = \frac{-3kT}{a^2}(-X_0 + 2X_1 - X_2)$$

$$f_{ix} = \frac{-3kT}{a^2}(-X_{i-1} + 2X_i - X_{i+1}) \qquad 1 \leqslant i \leqslant z-1 \tag{7-42}$$

$$f_{zx} = \frac{-3kT}{a^2}(-X_{z-1} + X_z)$$

In equation (7-42) f_{ix} is the force on the ith bead in the x direction, X_i is the amount by which bead i has been displaced from its equilibrium position, and a^2 is the mean square end-to-end distance of the submolecule. The form of equation (7-42) results from the fact that the x directed force on the ith bead reflects the difference between the x directed forces on the ith and $i+1$st segments. An additional force acts on the molecule due to the viscous nature of the medium in which it is immersed. Under the assumption that the beads move in the manner of spheres through a viscous solvent, the drag force on each bead is

$$f_{ix} = \rho \frac{dX_i}{dt} = \rho \dot{X}_i \qquad (7\text{-}43)$$

where ρ is the segmental friction factor. Under the further assumption that the forces arising from the acceleration of the beads are small, the elastic forces given in equation (7-42) and the viscous forces given in equation (7-43) must balance. This allows us to write:

$$\rho \dot{X}_0 = \frac{-3kT}{a^2}(X_0 - X_1)$$

$$\rho \dot{X}_i = \frac{-3kT}{a^2}(-X_{i-1} + 2X_i - X_{i+1}) \qquad 1 \leqslant i \leqslant z - 1 \qquad (7\text{-}44)$$

$$\rho \dot{X}_z = \frac{-3kT}{a^2}(-X_{z-1} + X_z)$$

This set of linear first-order differential equations may be represented in matrix notation as:

$$[\dot{X}] = -B[A][X]$$

where

$$[\dot{X}] = \begin{bmatrix} \dot{X}_0 \\ \dot{X}_1 \\ \vdots \\ \dot{X}_z \end{bmatrix} \qquad [X] = \begin{bmatrix} X_0 \\ X_1 \\ \vdots \\ X_z \end{bmatrix}$$

[A] is the square $(z+1)\times(z+1)$ matrix:

$$[A]=\begin{bmatrix} 1 & -1 & 0 & \cdot & \cdot & & \cdot & & \cdot \\ -1 & 2 & -1 & 0 & \cdot & & \cdot & & \cdot \\ 0 & -1 & 2 & -1 & 0 & & \cdot & & \cdot \\ 0 & 0 & \cdot & \cdot & \cdot & & \cdot & & \cdot \\ 0 & 0 & \cdot & \cdot & \cdot & -1 & 2 & -1 \\ \cdot & \cdot & \cdot & \cdot & \cdot & 0 & -1 & 1 \end{bmatrix}$$

and B is the composite constant $3kT/a^2\rho$.

Equation (7-44) represents the total physical content of our model. From here on, we will be concerned with mathematical methods that allow us to solve this apparently complicated set of equations. The main problem arising in any attempted solution of these differential equations results from coupling of the motions of the beads. Thus X_i is not a function of the position of the ith bead itself, but is directly dependent on the position of the adjacent beads. This rather standard problem is effectively treated using the techniques of normal coordinates. We will define a new set of coordinates, q_i, made up of a linear combination of the X_i's. Thus these new coordinates will be defined as

$$q_i = \sum_j Q_{ij}X_j \tag{7-45}$$

or, in matrix notation:

$$[q]=[Q][X]$$

where $[q]$ and $[X]$ are column matrices and $[Q]$ is a $(z+1)$ by $(z+1)$ square array. It remains for us to define $[Q]$ in such a way that it will lead to the solution of equation (7-44). To do this, we wish to be able to express every equation in the set equation (7-44) in terms of the normal coordinate and its time derivative alone. In matrix notation we formally write:

$$[\dot{q}]=-B[\Lambda][q] \tag{7-46}$$

where $[\Lambda]$, unlike $[A]$, is diagonal. Now, for example, the jth equation in the set of equation (7-46) reads

$$\dot{q}_j = -B\lambda_j q_j \tag{7-47}$$

Since only the jth normal coordinate and its time derivative appear in equation (7-47), direct integration yields the time dependence of the motion of this coordinate. Clearly, it is in general difficult to have any "feel" for what motion each normal coordinate represents since it has a complicated dependence on *all* of the real coordinates. Nevertheless, the sum of the motions of all of the normal coordinates is identically equal to the sum of the motions of all of the real coordinates since one is just a linear transform of the other [equation (7-45)].

Thus, to proceed mathematically, we must transform equation (7-44) into equation (7-46), which is done by diagonalizing the matrix $[A]$. There exists another matrix $[Q]$ such that

$$[Q^{-1}][A][Q] = [\Lambda] \tag{7-48}$$

Also

$$[Q^{-1}][Q] = [I] \tag{7-49}$$

In equation (6-37) $[I]$ is a diagonal matrix with all nonzero elements $= 1$. Operation by $[Q^{-1}]$ from the left on equation (7-44) leads to

$$[Q^{-1}][\dot{X}] = -B[Q^{-1}][A][Q][Q^{-1}][X] \tag{7-50}$$

If we now set

$$[Q^{-1}][\dot{X}] = [\dot{q}] \quad \text{and} \quad [Q^{-1}][X] = [q] \tag{7.51}$$

our problem is solved, since equation (7-50) becomes

$$[\dot{q}] = -B[\Lambda][q] \tag{7-52}$$

It must be reemphasized that the exact nature of $[Q^{-1}]$ is not necessary to the physical solution of our problem. Since the normal coordinate approach merely represents a linear transformation of the real coordinates, the motion of the polymer represented by all the q_i's will be identical to the motion of the polymer represented by all the x_i's. Our problem thus becomes the rather simple one of finding a diagonal representation of the $(z+1) \times (z+1)$ matrix $[A]$.[†] This rather well known result (this matrix applies in the treatment of a vibrating string, among others) is derived in

[†]Note that $[A]$ was a $(z+1) \times (z+1)$ square matrix while $[\Lambda]$ is a $z \times z$. This difference in order represents the fact that while translation of the total molecule is expressed in $[A]$, it is not built into $[\Lambda]$. This causes no complication.

the appendix at the end of this chapter, and is merely stated here:

$$\lambda_p = 4\sin^2\left(\frac{p\pi}{2(z+1)}\right) \tag{7-53}$$

where p goes from 1 to z. Thus the diagonal matrix $[\Lambda]$ is

$$[\Lambda] = \begin{bmatrix} \lambda_1 & 0 & 0 & \cdot & & \cdot & \cdot \\ 0 & \lambda_2 & 0 & \cdot & & \cdot & \cdot \\ 0 & 0 & \lambda_3 & \cdot & & \cdot & \cdot \\ \cdot & \cdot & \cdot & \lambda_{z-1} & & 0 & \\ \cdot & \cdot & \cdot & & 0 & & \lambda_z \end{bmatrix} \tag{7-54}$$

and

$$\dot{q}_p = -B\lambda_p q_p \tag{7-55}$$

which can be directly integrated to give

$$q_p(t) = q_p(0)e^{-Bt\lambda_p} = q_p(0)e^{-t/\tau_p}$$

$$\tau_p = \frac{1}{B\lambda_p} \qquad p = 1 \text{ to } z \tag{7-56}$$

Here $q_p(0)$ is the value of the normal coordinate at time zero, that is, at the application of a perturbation, and $q(t)$ is the value of the coordinate at time t. We see that the coordinate response is exponential and the system response is just the sum total of all the coordinate responses.

As mentioned above, however, the exact nature of the normal coordinate in terms of the real coordinates is not easily visualizable. Likewise, the exact nature of the real perturbation is not easily visualizable in terms of perturbation to the normal coordinates. Thus to carry out our calculation exactly, one would have to transform the perturbation into the normal coordinate framework. This is exactly the technique used by Bueche in reference 3. The perturbation, that is, the boundary condition, used to solve equation (7-55) was that every normal coordinate was instantaneously displaced to the position of $q_i(0)$ at time zero and then no additional forces were put on the system. This perturbation corresponds neither to creep nor to stress relaxation. Although boundary conditions corresponding to these real experiments are more complicated in terms of normal coordinates,[3] it can be shown[2] that the relaxation times that arise in a stress-relaxation experiment are just one-half as large as those calculated above. Thus, from here on τ_p will be used to denote a relaxation time and is given by

$$\tau_p = \frac{1}{2B\lambda_p} \tag{7-57}$$

Now one may associate those relaxation times with the relaxation times of a Maxwell–Wiechert model. Thus the stress relaxation behavior for the bead-and-spring model is given as

$$E(t) = \sum_{p=1}^{z} E_p e^{-t/\tau_p} \tag{7-58}$$

where the individual τ_p's are given in equation (7-57).

It remains to define the individual E_i's; here one normally relies on the kinetic theory of rubber elasticity as applied to the submolecules. Once again,

$$f = \frac{3kT}{a^2} \Delta x \tag{7-42}$$

is the elastic force experienced by each spring if the ends are perturbed some amount Δx. Consider, however, that we have N polymer molecules per cm^3 and the average cross-sectional area of each is b^2. Thus the stress experienced by each spring should be given as

$$\sigma = \frac{f}{b^2} = \frac{3kT}{b^2 a^2} \Delta x \tag{7-59}$$

The instantaneous tensile modulus is just the stress divided by the strain, so

$$E(0) = \frac{\sigma}{\varepsilon} = \frac{3kT}{b^2 a} \tag{7-60}$$

The denominator of the right side of equation (7-60) is merely the volume occupied per submolecule, given as

$$ab^2 = \frac{1}{cz} \tag{7-61}$$

Here c is the polymer concentration in molecules per cubic centimeter. Substitution of equation (7-61) into (7-60) yields

$$E(0) = 3ckTz \tag{7-62}$$

Clearly, however, the short time-limit of equation (7-58) is just

$$E(0) = \sum_{p=1}^{z} E_p \qquad (7\text{-}63)$$

Now one usually assumes that all of the individual E_p's are equal and given as

$$E_p = 3ckT \qquad (7\text{-}64)$$

so that equations (7-62) and (7-63) are consistent. Since we have the τ_p's and E_p's of the Maxwell–Wiechert model we may immediately write, for example,

$$E'(\omega) = 3ckT \sum_{p=1}^{z} \frac{\omega^2 \tau_p^2}{1 + \omega^2 \tau_p^2} \qquad (7\text{-}65)$$

D. APPLICATIONS OF FLEXIBLE CHAIN MODELS TO SOLUTIONS

Since we are primarily concerned with the viscoelastic nature of bulk materials in this introductory text, we will not dwell on the application of these ideas to solutions but, considering that the model has been derived for the case of a dilute solution, it is of interest to briefly examine its agreement with the observed viscoelastic response of solutions. First it must be recognized that neither a^2, the mean square end-to-end distance of the submolecule nor ρ, the segmental friction factor, may be easily evaluated. Thus a method of eliminating them from our equations will be helpful. The solution viscosity in excess of the solvent viscosity can be thought of as arising from the dissolved polymer and since our polymer behavior has been cast into the Maxwell–Wiechert framework, this excess viscosity is just the sum of the viscosity of each of the elements in the model.

$$\eta - \eta_s = \sum_{p=1}^{z} G_p \tau_p \qquad (7\text{-}66)$$

where η is the shear viscosity of the solution, η_s is the shear viscosity of the solvent, and G_p is one third of E_p. Combining equation (7-66) with (7-64),

(7-57), and (7-53) yields

$$\eta - \eta_s = ckT \sum_{p=1}^{z} \frac{1}{2B\lambda_p} = \frac{ca^2\rho}{24} \sum_{p=1}^{z} \frac{1}{\sin^2\{p\pi/[2(z+1)]\}} \tag{7-67}$$

Now for small values of X,

$$\sin X \approx X \tag{7-68}$$

and since this expression is in the denominator of each term in the sum in equation (7-67), where the smallest arguments of the sine function will contribute most, we may write

$$\sum_{p=1}^{z} \frac{1}{\sin^2\{p\pi/[2(z+1)]\}} \approx \sum_{p=1}^{z} \frac{4(z+1)^2}{p^2\pi^2} = \frac{4(z+1)^2}{\pi^2} \sum_{p=1}^{z} \frac{1}{p^2} \tag{7-69}$$

This summation is well known, and for large values of z it is equal to $\pi^2/6$. Furthermore, for large values of z, 1 is small compared to z so that substitution of equation (7-69) into equation (7-67) yields

$$\eta - \eta_s \approx \frac{ca^2\rho z^2}{36} \tag{7-70}$$

Applying the same arguments to equation (7-57) yields

$$\tau_p = \frac{a^2\rho z^2}{6\pi^2kTp^2} \qquad \text{for} \quad p < z \tag{7-71}$$

Combination of equation (7-70) and equation (7-71) gives

$$\tau_p = \frac{6(\eta - \eta_s)}{ckT\pi^2 p^2} \tag{7-72}$$

where the excess viscosity and concentration of the solution are known or are easily measured.

Rouse and Sittel[5] have investigated the applicability of the theory to real systems, in particular, dilute solutions of polystyrene in the good solvent toluene. Their results are reproduced in Figure 7-16. The agreement between theory and experiment is excellent. However, in a sense a certain amount of "curve fitting" is involved, since the friction factors and a^2 have

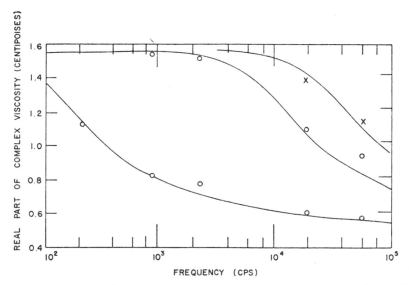

Figure 7-16. Experimental observation and theoretical prediction of the Rouse theory. After J. D. Ferry, *Viscoelastic Properties of Polymers*, Figure 10-4, by permission of John Wiley & Sons, Inc.

been adjusted to fit the data through the method outlined in deriving equation (7-72).

E.   THE ZIMM MODIFICATION

Consideration of another major modification that has been applied to the flexible chain model seems pertinent at this point. It has long been appreciated that the velocity gradient of the solvent would be perturbed deep inside a coiled polymer molecule. It is clear that this effect is not considered in the above treatment since the viscous drag is given as $\rho \dot{X}_i$ in equation (7-43) irrespective of whether X_i happens to be inside the coiled molecule or on its surface. Thus one might expect the Rouse formulation to be most applicable to polymer–solvent systems that result in elongated conformations of polymer chains predominating. In such a conformation, little shielding would result from the velocity gradient effective on one part of a molecule by another part of the molecule. This is the case in Figure 7-16 where toluene, a good solvent for polystyrene, would favor extended conformations. Balanced against this is the assumption of the model that the Gaussian distribution function holds. This is not true in a solvent that favors extended conformations and renders the agreement shown in Figure 7-16 probably fortuitous.

Zimm[4] has developed a theory that treats such "hydrodynamic shielding" and although we will not go into detail, it is helpful to examine the results of this calculation. The main difference between the Rouse and Zimm treatments occurs in the relaxation times, not in the partial modulus values. The relaxation times according to the Zimm treatment are

$$\tau_p = \frac{1.71(\eta - \eta_s)}{ckTK_p} \tag{7-73}$$

where the K_p are constants; $K_1 = 4.04$; $K_2 = 12.79$; $K_3 = 24.2$, and so on. In Figure 7-17 we have compared the prediction of the Rouse and Zimm theories.[6] The limiting values of the slopes of plots of log G' versus log ω at large ω are one-half for the Rouse treatment and two-thirds for the Zimm treatment. This arises from the different distributions of relaxation times in the Rouse and Zimm theories. In Figure 7-18 the results of DeMallie,[7] on the system of polystyrene in arochlor are presented. Arochlor, a rather poor solvent for polystyrene, yields a predominance of tightly coiled polymer conformations where hydrodynamic shielding would be expected to be most important. The remarkable agreement between theory and experiment

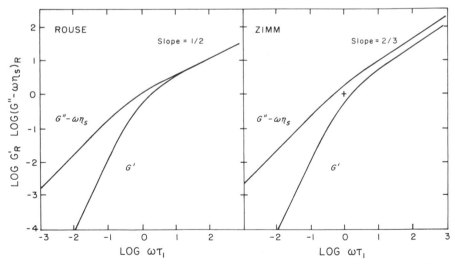

Figure 7-17. Predictions of the Rouse and Zimm theories. After J. D. Ferry et al., *J. Phys. Chem.*, **66**, 536 (1962). Reprinted by permission of the American Chemical Society.

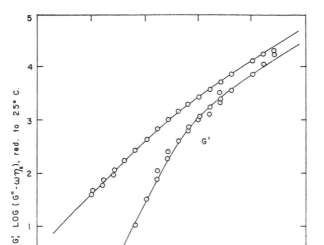

Figure 7-18. Experimental results for polystyrene in arochlor compared with predictions according to the Zimm theory. After J. D. Ferry et al., *J. Phys. Chem.*, **66**, 536 (1962). Reprinted by permission of the American Chemical Society.

in Figures 7-16 and 7-18 illustrate the validity of the bead-and-spring model for the viscoelastic behavior of polymer solutions.

F. EXTENSION TO BULK POLYMER

Returning to our major concern, bulk polymeric behavior, we may ask if we should expect the bead-and-spring model to be applicable at all. The equations of motion, equation (7-44) were written on the premise that the beads encountered viscous drag by virtue of their immersion in a solvent with viscosity η. No solvent is present in the bulk case, and it would thus appear that equation (7-43) is not applicable to this situation. However, the functional form of the viscous drag term [the left side of equation (7-44)] merely states that the solvent medium exerts the same drag on the bead regardless of the bead's position. This functional form can be preserved by substituting an ensemble of bead–spring polymer molecules for the solvent. This procedure has the effect of changing the numerical value of the friction factor, ρ, but retaining the left side of equation (7-44) unchanged.

The right side of equation (7-44), representing the restoring force resulting from the perturbation of a submolecule, is an intramolecular property, at least to a first approximation, and thus would not be expected to change drastically with the substitution of additional polymer for solvent. Therefore equation (7-44) is still applicable to the bulk polymer with the appropriate changes in numerical values for such parameters as ρ and a^2. The usual normal coordinate analysis leads directly to equation (7-57) for the distribution of the z relaxation times. The partial modulus now is NkT as opposed to ckT where N is the number of polymer chains per cubic centimeter of bulk polymeric material. An analysis equivalent to that given in equations (7-66) through (7-72) yields the relationship:

$$\tau_p = \frac{6\eta}{NkT\pi^2 p^2} \tag{7-74}$$

where η now represents the steady-flow shear viscosity of a bulk polymer. The time-dependent shear stress relaxation modulus would thus be given as

$$G(t) = NkT \sum_{p=1}^{z} e^{-t/\tau_p} \tag{7-75}$$

A plot of $\log G(t)$ versus $\log(t/\tau_1)$ is shown in Figure 7-19. We have assumed that $N = 6 \times 10^{18}$ molecule/cm^3 which corresponds roughly to a molecular weight of 10^5 and a density of 1 g/cm^3; NkT, the modulus associated with each relaxation time is then 3.2×10^4 Pa at 385°K. The number of terms in the summation of equation (7-75), or equivalently, the number of submolecules in the polymer molecule, z, is not determined. We have plotted $G(t)$ for several values of z. It should be clear that $z = 1$ is just the Maxwell body behavior.

Comparison of the curves in this figure with those presented in Chapter 3 giving the oberved experimental response of bulk polymers is very disappointing. Experimentally, high-molecular-weight polymers exhibit two major transitions while the bead-and-spring model predicts only one.

Ferry, Landel, and Williams[8] have suggested a simple but appealing explanation for this apparent shortcoming of the thoery and, moreover, have put forth a modification of the theory that "predicts" two transitions as observed experimentally. In expounding their treatment, let us revert to some of the considerations regarding the molecular mechanisms giving rise to these transitions as explained in Chapter 3, Section A. There we deduced that the behavior in the rubbery region was due mainly to the long-range translational motions of an entangled rubbery mass. The molecular motion responsible for the behavior observed in the primary transition region,

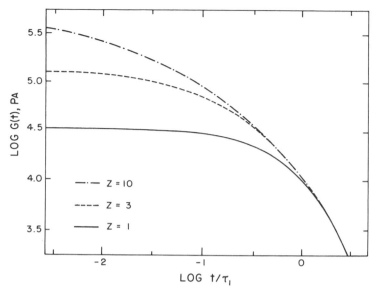

Figure 7-19. Rouse theory in the rubbery plateau and flow regions. Varying numbers of segments considered.

however, was argued to be of a much shorter range, involving perhaps only a few monometer units. Consider now the segmental friction factor ρ used in equation (7-44). If the motion of submolecule j is part of a coordinated long-range type of motion, we would expect the friction factor to reflect the entangled nature of the system, since for a long-range motion to take place entanglements must be affected. Conversely, in short-range motions, entanglements should play only a diminishingly small role; the friction factor would be expected to be indicative of the same chemical species without entanglements (low-molecular-weight polymer). Thus Ferry, Landel, and Williams argued that one should expect two independent friction factors to be operative. One of these reflects the viscous drag experienced by a submolecule taking part in short-range uncoordinated motions with short relaxation times, and the other is indicative of the viscous drag experienced by a submolecule taking part in long-range coordinated motions with long relaxation times.

We now ask which of the two friction factors has been used in the expression for τ_p equation (7-74). ρ has been eliminated from this expression through its dependence on η, the steady-state viscosity. Our question becomes, then, what molecular motion does η depend upon? Quite possibly it could reflect all motions of the polymer molecules, but fortunately for

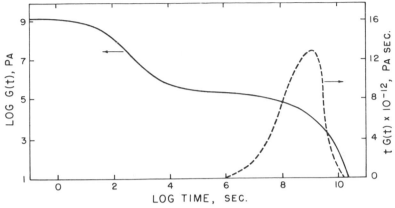

Figure 7-20. Calculations of the viscosity from a stress relaxation master curve via equation (3-7).

this discussion, it is rather simple to show that only long-range coordinated motions contributed substantially to η. To do this, recall equation (3-7):

$$\eta = \int_0^\infty G(t)\,dt \tag{3-7}$$

where we are now calculating a shear viscosity. (See problem 7-8 for the development of this equation.) A simple transformation yields

$$\eta = \int_{-\infty}^\infty tG(t)\,d\ln t \tag{7-76}$$

In Figure 7-20 is plotted the integrand of this integral along with a typically reduced master curve of a high-molecular-weight polymer. It is evident that the only significant contribution to the area comprising η comes from the long-time portion of the master curve, clearly indicating that the above statement concerning long-range motions as being the sole contribution to η is correct.

The friction factor operative for the short-range motions may be thought of as being obtained from measurements on polymers where entanglements are not present and the only contribution to all physical properties, η included, comes from short-range motions. Such a polymer is just one of low molecular weight where entanglements are not possible because the chains are too short. Polymers with a critical length for the onset of entanglement have been found to have a molecular weight of about 30,000 (polystyrene). Thus Ferry, Landel, and Williams postulate one friction factor ρ_0, operative at relaxation times shorter than some critical relaxation time τ_c and another, ρ, operative at longer relaxation times:

$$\tau_p = \frac{\rho_0 a^2 z^2}{6\pi^2 kTp^2} \qquad \tau_p < \tau_c$$

$$\tau_p = \frac{\rho a^2 z^2}{6\pi^2 kTp^2} \qquad \tau_p \geq \tau_c$$

(7-77)

Moreover, they postulate a relationship ρ/ρ_0 as

$$\log \frac{\rho}{\rho_0} = 2.4 \log \frac{M}{M_c} \tag{7-78}$$

where M_c is the critical molecular weight for the onset of entanglement. We see that they postulate a friction factor ρ which varies as molecular weight to the 2.4 power, unlike ρ_0 which is not a function of molecular weight. The distribution of relaxation times given in equation (7-77) was used to construct the relaxation master curve shown in Figure 7-21, where τ_c was taken to be $\tau_1/400$ and NkT as 1×10^4 Pa with arbitrary ρ/ρ_0. This curve is

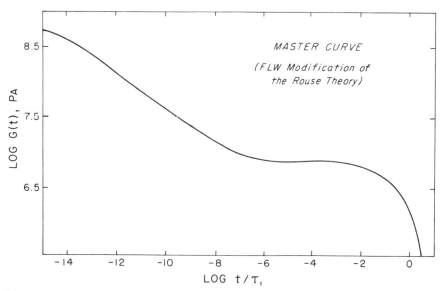

Figure 7-21. Two transitions predicted using the Ferry, Landel, and Williams[8] modification of the Rouse theory.

presented merely to demonstrate that the Ferry, Landel, and Williams modification to flexible chain theories does, in fact, predict in a straightforward way a two-transition relaxation curve. With a judicious choice of τ_c it is possible to reproduce the gross characteristics of most experimental master curves.

Remembering that the viscosity of a Maxwell–Wiechert model is just the sum of the viscosities of the individual Maxwell elements will help shed light on the 2.4 factor in equation (7-78). We have already noted that all of the viscosity of a high-molecular-weight polymer derives from long-range translational motion of the polymer, that is, from motions with $\tau_p > \tau_c$ or where the friction factor ρ is operative. Thus we may write

$$\eta = \sum_{p=1}^{z} G_p \tau_p \qquad \tau_p > \tau_c \tag{7-79}$$

and

$$\eta = \frac{NkT\rho a^2 z^2}{6\pi^2 kT} \sum_{p=1}^{P_c} \frac{1}{p^2} \tag{7-80}$$

where p_c is defined as

$$p_c = \sqrt{\frac{\tau_1}{\tau_c}} \tag{7-81}$$

If p_c is greater than five, for example, the summation is relatively insensitive to changes in p_c and essentially may be considered as a constant. Examination of each term in equation (7-80) with respect to molecular-weight dependence reveals

$$\eta \propto M^{3.4} \tag{7-82}$$

since N, the number of chains per cubic centimeter is inversely proportional to molecular weight while z, the number of submolecules per polymer molecule, is directly proportional. ρ, it has been pointed out, varies as the 2.4 power of molecular weight while all the other terms are independent of molecular weight. Realizing that equation (7-82) is an experimentally observed fact for high-molecular-weight materials makes clear the choice of 2.4 in equation (7-78).

In order to keep this subject in perspective, several shortcomings of the Ferry, Landel, and Williams modification of the Rouse theory must be kept in mind. Although these considerations indicate that the modification does not strictly account for all observations, it must be appreciated that this was a major step in making a degree of theoretical sense out of a large amount of sometimes confusing experimental data.

The argument of Williams[9] is perhaps the most serious blow to the Ferry–Landel–Williams work. He examined the short time-limit of equation (7-75) and observed that

$$G(0) = NkT \sum_{p=1}^{z} (1) = zNkT \qquad (7\text{-}83)$$

He pointed out that this relationship is true no matter what value of the friction factor is used since the exponential of $(0/\tau_p)$ is always 1.0. The limiting value of the modulus at short times, which is experimentally found to be in the neighborhood of 3×10^9 Pa for most polymers (Chapter 3), is clearly predicted from equation (7-83). Consider a polymer of molecular weight 150,000 and a density of 1.5 g/cm^3. The value for NkT at 300°K is 2.5×10^4 Pa.

What, then, is the value of z? One should recall that a submolecule was defined as the shortest unit of a chain whose end-to-end distance is Gaussian. Clearly, this would not be true of a monomeric or dimeric submolecule due to internal constraints on molecular geometry. Williams decided that the smallest unit he could accept for a submolecule was composed of five monomer units. If we let our polymer be polystyrene, whose monomer molecular weight is 100, we have 1500 monomer units per chain, or 300 submolecules per polymer chain. Using equation (7-83)

$$G(0) = (2.5 \times 10^4 \text{ Pa})(300) = 7.5 \times 10^6 \text{ Pa} \qquad (7\text{-}84)$$

which is about two orders of magnitude smaller than the experimentally determined value. The basis for this shortcoming seems to lie in the fact that one uses NkT as a partial modulus for all chain motion. This value, while clearly applicable to rubberlike systems, is not so clearly applicable to deformations of the glassy state. Tobolsky[10] has pointed out that the equations of motion [equation (7-43)] can be written and solved for chain units much smaller than the submolecule while employing a partial modulus more realistic for the types of molecular motion expected in the glassy state. Thus a Ferry–Landel–Williams type of modification is still feasible in spite of Williams' objection.

Another major discrepancy between theory and experiment is exemplified in Figure 7-22. In this figure, the predicted relaxation curves according

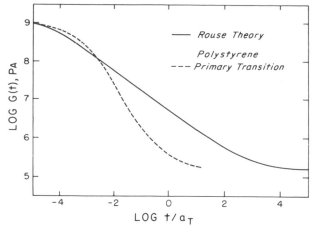

Figure 7-22. Comparison between master curve of polystyrene and predictions of the modified Rouse theory in the primary transition region.

to the Rouse theory are compared with the experimental results for polystyrene in the primary transition region. It is clear that polystyrene undergoes its glass-to-rubber transition much more sharply than is predicted by flexible chain theories. Many other polymers, in contrast to this, obey the flexible-chain-theory result rather closely. Polyisobutylene is one example of this. The discrepancy noted in the case of polystyrene is important, however, as we shall show. The question arises, What molecular factors contribute to the discrepancy in the case of polystyrene? In order to explore this, we use the maximum slope of the stress relaxation master curve in the glass-to-rubber transition region as a criterion for agreement between theory and experiment. Values of this slope measured experimentally for several polymers are listed in Table 7-3.

The theoretical expression for the modulus is taken from equation (7-75) with the relaxation times given in equation (7-77). Here $\tau_p < \tau_c$ since we are

Table 7-3. Transition Region Slope for Several Polymers[12]

Polymer	Slope
Polyvinyl chloride	−0.6
Polyisobutylene	−0.7
Polyethylene terephthalate	−0.9
Polycarbonate	−1.2
Polystyrene	−1.5

concerned about the transition region where short-range motions dominate. Let us define a fictitious relaxation time τ_l such that

$$\tau_l = \frac{\rho_0 a^2 z^2}{6\pi^2 kT}$$

(7-85)

where τ_l would be the maximum relaxation time τ_1 if one were not forced to change friction factors from ρ_0 to ρ for the long relaxation times. For all $\tau_p < \tau_c$ one may then write

$$\tau_p = \frac{\tau_l}{p^2}$$

(7-86)

and equation (7-75) is modified to become

$$G(t) = G \sum_{p=p_c}^{z} e^{-tp^2/\tau_l}$$

(7-87)

where G, the partial modulus associated with each relaxation, may or may not be NkT but is the same for all short-range molecular motions. Similarly, z is the number of molecular units considered per chain and is not necessarily the usual number of submolecules. It is argued now that for relatively large values of p, say $p > 5$, the summation may be replaced by an integration to yield

$$\frac{G(t)}{G} = \int_{p=p_c}^{z} e^{-tp^2/\tau_l} \, dp$$

(7-88)

Letting

$$X = \frac{tp^2}{\tau_l}$$

(7-89)

$$\frac{G(t)}{G} = 1/2 \left(\frac{\tau_l}{t}\right)^{1/2} \int_{t/\tau_c}^{t/\tau_{min}} X^{-1/2} e^{-X} \, dX$$

where $\tau_{min} = \tau_l/z^2$. Although this integral cannot be carried out analytically, it is the well-known incomplete gamma function that has been tabulated:[13,14]

$$\Gamma_b(a) = \int_0^b x^{a-1} e^{-x} \, dx$$

(7-90)

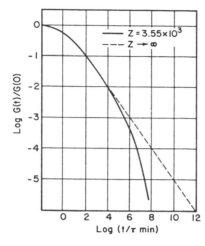

Figure 7-23. Relaxation in the primary transition region considering varying numbers of relaxing elements.

Thus:

$$\frac{G(t)}{G} = 1/2\left(\frac{\tau_l}{t}\right)^{1/2}\left[\Gamma_{t/\tau_{\min}}(1/2) - \Gamma_{t/\tau_c}(1/2)\right] \qquad (7\text{-}91)$$

The function $G(t)/G(0)$ has been plotted in Figure 7-23 and for times between $\tau_{\min}$ and τ_c it is clear that the slope of $\log G(t)$ versus $\log t$ is $-\frac{1}{2}$; this value arises from the distribution of relaxation times varying as $1/p^2$, the result of the normal mode analysis. Thus we see that this inflexibility of the theory is based in some of its very fundamental aspects.

Tobolsky[11] has suggested that the response of a highly coupled system would be more like that experimentally observed for polystyrene and has suggested ways to solve the equations generated from these considerations. Instead of considering the response of a linear chain, he treats a network coupled, via springs, in two and three dimensions. This treatment is considered beyond the scope of an introductory book; interested readers are referred to the original papers.[10–12,15]

G. REPTATION

Although theories of the Rouse–Bueche–Zimm type have been very successful in rationalizing the behavior of polymeric systems from a molecular point of view, another class of theories is presently under development. These theories, which treat the motion of polymer molecules in terms of reptation ("creeping"), that is, in terms of a wormlike motion of one polymer molecule through the matrix formed by its neighbors, are of particular interest since they are designed to avoid some of the shortcomings of the normal mode theories. We will first briefly sketch some of these shortcomings.

As mentioned above, it is not clear that the concept of "ideal chain elasticity" is an appropriate assumption for modeling the dynamics of high-molecular-weight polymeric solids. In equation (7-84), for example, it is clear that the maximum value of the modulus predicted by the normal mode theories is much smaller than that observed experimentally. And even in solution, it may not be correct to use the spring constant associated with an ideal chain.[16] Also, the complications arising from hydrodynamic screening, which Zimm has considered for the case of dilute solutions, have been essentially ignored. Finally, in the normal mode theories no direct account is taken of interchain interactions; it is tacitly assumed that one chain may pass through another chain to execute its normal-mode motion as if the latter were not present, that is, as if it were a phantom chain (Chapter 6, Section B1). Clearly, the influences of entanglements become difficult to visualize in this picture.

The reptation model, on the other hand, recognizes such impediments to chain motion. In Figure 7-24a a single chain is shown along with randomly

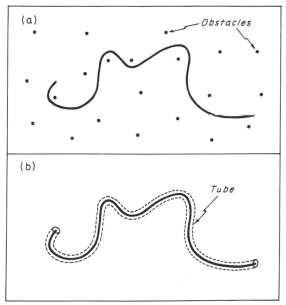

Figure 7-24. Reptation model view of a polymer chaim: (a) with obstacles; (b) in a tube.

placed dots that represent fixed obstacles to the chain motion. The reptation model suggests that the chain must move through this obstacle course in a wormlike fashion as relaxation occurs.

Exact calculations based on this model are complex. Nevertheless, it is relatively easy to develop certain "scaling laws" that relate how various macroscopic properties might depend on molecular properties. We will briefly sketch the development of such a scaling law for viscosity and chain length (or molecular weight) based on the reptation model.[16]

For convenience, motion of the chain through the set of obstacles of Figure 7-24a may be thought of as the motion of a chain constrained to move in a tube as shown in Figure 7-24b. The maximum relaxation time, τ_{max}, can be associated with the time necessary for the molecule to diffuse out of the constraining tube. This is so since the contour of the original tube would be influenced by external stresses applied to the system, and these would be totally relaxed when the molecule was no longer contained in the tube. Thus, the problem becomes one of calculating diffusion times of polymer molecules in tubes.

According to De Gennes,[16] this may be done as follows. First, apply a steady force f to the chain and observe its velocity v in the tube. Under these circumstances the mobility of the molecule in the tube μ_{tube} is defined as

$$v = \mu_{tube} f \qquad (7\text{-}92)$$

Let n be the number of monomer units per molecule. To obtain the *same* velocity v_c with molecules of various lengths, that is with various values of n, the force must be directly proportional to n. Thus, equation (7.92) may be rewritten:

$$\frac{v_c}{\mu_{tube}} = f \propto n \quad \text{or} \quad \mu_{tube} = \frac{\mu_1}{n} \qquad (7\text{-}93)$$

where μ_1 is independent of chain length. Knowing the molecular mobility allows one to calculate the diffusion constant through the Nernst–Einstein equation[17]:

$$D_{tube} = kT\mu_{tube} = \frac{kT\mu_1}{n} = \frac{D_1}{n} \qquad (7\text{-}94)$$

where k is Boltzmann's constant and where D_1 is also independent of chain length. From the study of diffusion, it is well known that

$$D = \frac{x^2}{2t} \qquad (7\text{-}95)$$

(see, e.g., A. W. Adamson, *A Textbook of Physical Chemistry*, 2nd edition, Academic Press, New York, 1979, p. 60), where x is the average distance a molecule moves in time t in a medium of diffusion constant D. The direct combination of equations (7-94) and (7-95) gives

$$\tau_{max} \propto \frac{L^2}{D_{tube}} = \frac{L^2 n}{D_1} \propto n^3 \tag{7-96}$$

where the diffusion time has been identified with the maximum relaxation time and the diffusion distance with the tube length L. Clearly, the tube length is directly proportional to the polymer chain length and thus the maximum relaxation time is predicted to depend on chain length to the third power.

The viscosity associated with this reptating motion may now be calculated as

$$\eta = \tau E \tag{7-4}$$

In the reptation model, the elastic modulus E depends on distance between obstacles and is, therefore, not chain length dependent. Thus, the chain length dependence of η and τ_{max} are predicted to be the same, that is,

$$\eta \propto n^3 \tag{7-97}$$

Equation (7-97) is an example of a scaling law in that it indicates how viscosity should depend on or "scale with" chain length and, therefore, molecular weight. It is perhaps a bit surprising that this equation can be derived so simply considering the complicated picture of chain motion suggested by the reptation model. It is known from experiment that viscosity actually varies as the 3.4 power of molecular weight (equation 7-82).

No attempt at completeness has been made in this short discussion. Rather, we have tried to introduce the concept of molecular motion through reptation and to sketch, in a cursory fashion, the techniques used for developing scaling laws through this model.

APPENDIX

Equation (7-53), which will be derived in this appendix, represents the eigenvalues used in diagonalizing the matrix $[A]$ of equation (7-44)

$$[\dot{X}] = -B[A][X] \tag{7-44}$$

where $[A]$ is a square matrix of order $(z+1)$.

$$[A] = \begin{bmatrix} 1 & -1 & 0 & 0 & \cdots\cdots\cdots \\ -1 & 2 & -1 & 0 & \cdots\cdots\cdots \\ 0 & -1 & 2 & -1 & \cdots\cdots\cdots \\ \cdots\cdots\cdots\cdots & -1 & 2 & -1 \\ \cdots\cdots\cdots\cdots & 0 & -1 & 1 \end{bmatrix}$$

First transform $[A]$ into a form that will be easier to treat. To do this, note that

$$[A] = [C^T][C]$$

where $[C]$ is also a matrix of order $z \times (z+1)$ and has the form

$$[C] = \begin{bmatrix} 1 & -1 & 0 & 0 & \cdots\cdots\cdots \\ 0 & 1 & -1 & 0 & \cdots\cdots\cdots \\ 0 & 0 & 1 & -1 & \cdots\cdots\cdots \\ \cdots\cdots\cdots\cdots & 1 & -1 & 0 \\ \cdots\cdots\cdots\cdots & 0 & 1 & -1 \end{bmatrix}$$

and

$$[C^T] = \begin{bmatrix} 1 & 0 & 0 & 0 & \cdots\cdots\cdots \\ -1 & 1 & 0 & 0 & \cdots\cdots\cdots \\ 0 & -1 & 1 & 0 & \cdots\cdots\cdots \\ \cdots\cdots\cdots\cdots\cdots & & \\ \cdots\cdots\cdots\cdots & -1 & 1 \\ \cdots\cdots\cdots\cdots & 0 & -1 \end{bmatrix}$$

Thus, equation (7-44) can be written as

$$[\dot{X}] = -B[C^T][C][X] \tag{a}$$

Now premultiply both sides of equation (a) by $[C]$:

$$[C][\dot{X}] = -B[C][C^T][C][X]$$
$$= -B[R][C][X] \tag{b}$$

Where $[R]$ is called the Rouse matrix,

$$[R] = \begin{bmatrix} 2 & -1 & 0 & 0 & \cdots \cdots \cdots \cdots \\ -1 & 2 & -1 & 0 & \cdots \cdots \cdots \cdots \\ 0 & -1 & 2 & -1 & \cdots \cdots \cdots \cdots \\ \cdots \cdots \cdots \cdots & & & & \cdots \cdots \cdots \cdots \\ \cdots \cdots \cdots \cdots & -1 & 2 & -1 & 0 \\ \cdots \cdots \cdots \cdots & 0 & -1 & 2 & -1 \\ \cdots \cdots \cdots \cdots & 0 & 0 & -1 & 2 \end{bmatrix}$$

which is a $z \times z$ square matrix. (See footnote on page 162.) This matrix can now be diagonalized by the approach mentioned in the text:

$$[\varphi^{-1}][R][\varphi] = [\Lambda] \tag{c}$$

Operation by $[\varphi^{-1}]$ from the left on equation (b) gives

$$[\varphi^{-1}][C][\dot{X}] = -B[\varphi^{-1}][R][\varphi][\varphi^{-1}][C][X] \tag{d}$$

In equation (d) we took advantage of the fact that $[\varphi^{-1}][\varphi] = [I]$, given in equation (7-49). Now define normal coordinates

$$[q] = [\varphi^{-1}][C][X] \tag{e}$$

and

$$[\dot{q}] = [\varphi^{-1}][C][\dot{X}] \tag{f}$$

which are slightly different from equation (7-51). This is done strictly for mathematical convenience and has no consequence in the final solution. Equation (d) can thus be recast as

$$[\dot{q}] = -B[\Lambda][q] \tag{7-52}$$

where $[q]$ and $[\dot{q}]$ are column matrices and $[\Lambda]$ is a diagonal matrix.

Now diagonalize $[R]$ by rewriting equation (c) as

$$[R][\varphi] = [\varphi][\Lambda] \tag{g}$$

Equation (g) requires that the corresponding elements in the matrices on both sides must be equal after the multiplication has been carried out. Thus, for each value of Λ_{ii}, we obtain a set of linear equations as follows (the reader may wish to verify them for a 3×3 matrix):

$$(2-\lambda)\varphi_{11} - \varphi_{12} = 0$$

$$-\varphi_{21} + (2-\lambda)\varphi_{22} - \varphi_{23} = 0$$

$$-\varphi_{32} + (2-\lambda)\varphi_{33} - \varphi_{34} = 0 \qquad \text{(h)}$$

$$\cdots\cdots\cdots\cdots\cdots\cdots\cdots$$

$$-\varphi_{z,z-1} + (2-\lambda)\varphi_{zz} = 0$$

where φ_{ij} is the ijth term in $[\varphi]$.

In order to solve the set of equations given above, note that they all have the same form:

$$-\varphi_{m-1} + (2-\lambda)\varphi_m - \varphi_{m+1} = 0 \qquad \text{(i)}$$

with the condition that

$$\varphi_0 = \varphi_{z+1} = 0 \qquad \text{(j)}$$

Equation (i) can be treated and solved as a difference equation. In operator notation it is just

$$\left[-E^{-1} + (2-\lambda) - E \right]\varphi_m = 0 \qquad \text{(k)}$$

where the operator E displaces a function in the positive direction and E^{-1} in the negative direction, that is,

$$E\varphi_m = \varphi_{m+1}$$

$$E^{-1}\varphi_m = \varphi_{m-1}$$

Now to solve equation (k), assume that the solution has the following form:

$$\varphi_m = \beta e^{m\alpha} \qquad \text{(l)}$$

where β is a constant and α is a function to be determined. Insertion of equation (l) into (k) results in

$$\beta e^{m\alpha}\left[-e^{-\alpha} + (2-\lambda) - e^{\alpha} \right] = 0 \qquad \text{(m)}$$

In order for equation (m) to yield nontrivial results, it is necessary that

$$-e^{-\alpha}+(2-\lambda)-e^{\alpha}=0$$

or

$$\frac{2-\lambda}{2}=\frac{e^{\alpha}+e^{-\alpha}}{2}=\cosh\alpha \qquad (n)$$

Equation (n) is satisfied by two values of α, namely $+\alpha$ and $-\alpha$, since $\cosh(+\alpha)=\cosh(-\alpha)$. Thus, the solution to the difference equation also must be satisfied by two functions ($\beta e^{+m\alpha}$ and $\beta e^{-m\alpha}$). The general solution is then just

$$\varphi_m = P_1 \beta e^{m\alpha} + P_2 \beta e^{-m\alpha} \qquad (o)$$

where the P's are constants. Since

$$\sinh x = \frac{e^x - e^{-x}}{2}$$

and

$$\cosh x = \frac{e^x + e^{-x}}{2}$$

we can have the following equivalent general solution:

$$\varphi_m = M_1 \sinh(m\alpha) + M_2 \cosh(m\alpha) \qquad (p)$$

where the M's are constants. For $m=0$, equation (j), the boundary condition, demands

$$\varphi_0 = 0 = M_2 \qquad (q)$$

and

$$\varphi_{z+1} = 0 = \sinh(z+1) \qquad (r)$$

since M_1 is a constant. But the value of the hyperbolic sine is zero if the argument is an integral multiple of ($i\pi$), where $i=\sqrt{-1}$. Therefore,

$$\alpha = \frac{ip\pi}{z+1} \qquad p=1,2,3,\ldots,z \qquad (s)$$

Combination of equations (s) and (n) finally gives the desired result:

$$\lambda_p = 2 - 2\cosh\frac{ip\pi}{z-1}$$

$$= 2\left(1 - \cos\frac{p\pi}{z+1}\right)$$

$$= 4\sin^2\frac{p\pi}{2(z+1)} \tag{7-53}$$

where $p = 1, 2, 3, \ldots, z$.

PROBLEMS

1.   Calculate the slope of the curve in Figure 7-3b and its limiting value at times much greater than τ.

2.   Develop expressions for the complex modulus and compliance for a Maxwell body

   (a)   From the stress relaxation modulus of the body using the phenomenological theory of Chapter 2.

   (b)   From the equation of the Maxwell model equation (7-5) assuming a sinusoidal strain application.

3.   Develop expressions for D', D'', E', and E'' for a Voigt element

   (a)   From equation (7-25) using the phenomenological theory of Chapter 2.

   (b)   From the equation of motion of the Voigt element, equation (7-22), assuming a sinusoidal stress application. (Notice the appearance of a transient term when the boundary condition $\varepsilon(0) = 0$ is used.)

4.   Derive an expression for the complex modulus of a Maxwell–Wiechert model subjected to a sinusoidal strain. Show that the complex compliance is not obtainable as a simple analytical function.

5.   Derive an expression for the complex modulus of z Maxwell elements in series assuming a sinusoidal strain application.

6.   Bueche[3] has shown that the flexible bead-and-spring model leads to the following expression for the creep compliance of an undiluted polymer:

$$D(t) = \frac{8}{3NkT\pi^2}\sum\frac{1}{p^2}(1 - e^{-t/\tau_p})$$

where $p = 1, 3, 5, \ldots, z$.

In this case the retardation times are given by

$$\tau_p = \frac{\rho a^2 z^2}{3\pi^2 k T p^2} \qquad p = 1, 3, 5, \dots, z$$

Show that

(a)   The retardation times can be rewritten as

$$\tau_p = \frac{12\eta}{N\pi^2 k T p^2}$$

(b)   The partial compliances of the equivalent Voigt–Kelvin model are

$$D_p = \frac{8}{3NkT\pi^2 p^2}$$

7. Determine the parameters for a two-element Maxwell–Wiechert model (E_1, E_2, τ_1, τ_2) which would give a $\log E''(\omega)$ versus $\log \omega$ response similar to that shown below:

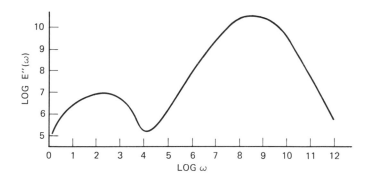

8. Show that equation (3-7) yields the viscosity for a Maxwell–Wiechert model.

9. Calculate the slope of a stress relaxation master curve at some time t when $\tau_{min} < t < \tau_{max}$ where

$$\tau_p = \frac{\tau_{max}}{p} \quad \text{and} \quad E_p = k$$

10. It is frequently necessary to obtain the function $H(\tau)$ from experimental data, and various approximations are useful in carrying out this transformation. Show that the so-called first approximation:

$$H_1(\tau) \equiv \frac{dE(t)}{dlnt}\bigg|_{t=\tau}$$

results from approximating the expression in equation (7-35) by a unit step at $t = \tau$.

11. (a)   Derive an expression for $H_1(\tau)$ for a Maxwell body. Plot this function as $\log H_1(\tau)/E_0$ versus $\tau/\tau_{Maxwell}$. Be sure to remember that the relaxation time for the Maxwell body is a constant.

   (b)   Compare the above result with $H_2(\tau)$, a second approximation of $H(\tau)$ which is given as

$$H_2(\tau) \equiv \left(-\frac{dE(t)}{dlnt} + \frac{d^2E(t)}{d(lnt)^2} \right)_{t=2\tau}$$

REFERENCES

1.   P. E. Rouse, *J. Chem. Phys.*, **21**, 1272 (1953).

2.   W. L. Peticolas, *Rubber Chem. Technol.* **36**, 1422 (1963).

3.   F. Bueche, *J. Chem. Phys.*, **22**, 603 (1954).

4.   B. H. Zimm, *J. Chem. Phys.*, **24**, 269 (1956).

5.   P. E. Rouse and K. Sittel, *J. Appl. Phys.*, **24**, 690 (1954).

6.   J. D. Ferry, *Viscoelastic Properties of Polymers*, 3rd ed., Wiley, New York, 1980, p. 213.

7.   R. B. DeMallie Jr., M. H. Birnboim, J.E. Frederick, N. W. Tschoegl, and J. D. Ferry, *J. Phys. Chem.*, **66**, 586 (1962).

8.   J. D. Ferry, R. F. Landel, and M. L. Williams, *J. Appl. Phys.*, **26**, 359 (1955).

9.   M. L. Williams, *J. Polym. Sci.*, **62**, 57 (1962).

10.   A. V. Tobolsky and D. B. DuPre, *Adv. Polym. Sci.*, **6**, 103 (1969).

11.   A. V. Tobolsky, *J. Chem. Phys.*, **37**, 1575 (1962).

12.   J. J. Aklonis, Ph.D. thesis, Princeton University, 1965.

13.   W. V. Houston, *Principles of Mathematical Physics*, 2nd ed., McGraw-Hill, New York, 1948.

14.   L. A. Pipes, *Applied Mathematics for Engineers and Physicists*, 2nd ed., McGraw-Hill, New York, 1958.

15.   M. Shen, W. F. Hall, and R. E. DeWames, *J. Macromol. Sci.*, **C2**, 183 (1968).

16.   P. G. De Gennes, *Scaling Concepts in Polymer Physics*, Cornell University Press, Ithaca, N.Y., 1979.

17.   T. Erdey-Gruz, *Transport Phenomena in Aqueous Solutions*, Wiley, New York, 1974.

18.   B. Gross, *Mathematical Structure of the Theories of Viscoelasticity*, Herman, Paris, 1953.

8

Dielectric Relaxation

When an insulating material is subjected to an applied electric field, charge separation and molecular rearrangement occur within the material, causing the phenomenon of polarization. The magnitude of the polarization is measured by a property of the material called the dielectric constant. This macroscopic property is in turn related to the molecular structure of the dielectric through the molecular polarizability and the molecular dipole moment. The exact form of this connection is far from obvious, and several theories have been proposed to explain it.

Dielectric relaxation, as the name implies, is concerned with the time, frequency, and/or temperature dependence of the dielectric constant. Since the magnitude of the dielectric constant is related to the molecular structure, its dependence on time, frequency, and/or temperature generally reflects molecular motion. In the case of polymers, the dielectric relaxation technique may therefore be used as a probe for the study of transitions and relaxations in a manner analogous to that already discussed for mechanical relaxation. In this chapter we are concerned with the application of dielectric relaxation to amorphous polymers, and we attempt to point out differences between the dielectric and mechanical relaxation techniques.

In a manner now familiar, we start by treating dielectric relaxation phenomenologically, that is, in macroscopic terms, where the existence of molecules is ignored. In Section B, our development is extended by incorporating molecular considerations. Applications of these ideas to polymers are treated in Sections C through F.

A. PHENOMENOLOGY

For the purposes of this discussion, we need only be concerned with electrical circuits that contain capacitances, C, and resistances, R. The resistance is the dissipative element, formally analogous to the dashpot in the mechanical model case. It is defined by Ohm's law:

$$R = \frac{V}{I} \tag{8-1}$$

where V is the voltage in volts, I is the current in amps, and R has units of ohms. A capacitor consists of two ideal electrodes separated by vacuum and is a conservative element, playing the same role as the spring in the mechanical model case. If a voltage V is applied to the capacitor plates in vacuum, the capacitor will hold a charge Q, which is measured in coulombs and which is related to the voltage as:

$$Q = C_0 V \tag{8-2}$$

C_0 is the capacitance and is measured in farads. If the plates are separated by a dielectric, that is, by an insulating material (rather than a vacuum), the capacitor will accept more charge at the same potential (due to polarization of the dielectric). Under these conditions, the capacitance becomes

$$C = \varepsilon C_0 \tag{8-3}$$

where ε is the dielectric constant. (The dielectric constant is alternately referred to as the relative permittivity.) As is clear from the definition, ε is a unitless quantity. For vacuum, $\varepsilon = 1$ by definition; $\varepsilon = 81$ for water, $\varepsilon \sim 6$ for various types of inorganic glasses, and $\varepsilon = 1.0006$ for air. Since ε is a function of time or frequency (as well as temperature) these values vary with experimental conditions. The quoted numbers are for low-frequency experiments (equilibrium values) at room temperature. The dielectric constant of water decreases to about 1.8 for frequencies involved in optical experiments ($\sim 10^{15}$ Hz).

We shall now show that a series combination of a capacitor and a resistance in an electrical circuit leads to the same linear differential equation describing the time dependence of the charge as that for the Voigt model describing the time dependence of the stress in the mechanical case.[1]

Figure 8-1 is a schematic representation of the circuit in question. For series electrical circuits, the voltage across the terminals is the sum of the

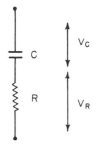

Figure 8-1. Schematic representation of a C–R circuit.

voltages across the elements:

$$V = V_C + V_R \qquad (8\text{-}4)$$

Now V_C is given by equation (8-2) and V_R by equation (8-1). The current is defined by

$$I = \frac{dQ}{dt} \qquad (8\text{-}5)$$

Substituting equation (8-5) into equation (8-4):

$$V = \frac{Q}{C} + R\frac{dQ}{dt} \qquad (8\text{-}6)$$

Now if we define

$$\tau = CR \qquad (8\text{-}7)$$

equation (8-6) may be rearranged to give

$$\tau\frac{dQ}{dt} + Q = CV \qquad (8\text{-}8)$$

This equation is entirely analogous to the equation of motion for the Voigt element in the mechanical case [equation (7-22)].

By comparing these two equations, it can be seen that the charge is the electrical analogue of the strain, as the voltage is the electrical analogue of the stress, and the capacitance corresponds to the mechanical compliance. It should be noted that a series electrical model gives rise to an equation of motion of the same form as a parallel mechanical model. This is because voltages add when in series while stresses add when in parallel.

Not surprisingly, equation (8-8) may be integrated for various boundary conditions in the same manner as the mechanical Voigt element described in Chapter 7. For example, for a constant voltage V_0, applied at time $t = 0$, the result is

$$Q(t) = CV_0(1 - e^{-t/\tau})$$ (8-9)

The integration of equation (8-8) is carried out using the integrating factor technique, which was fully explained in Chapter 7, equation (7-24). It should be clear that equation (8-9) represents the electrical analogue of the mechanical-creep experiment on a single Voigt element.

In the development so far, we have assumed that capacitance is independent of time, which is only strictly true for a vacuum. All real materials exhibit time-dependent capacitances, which arise from the time dependence of the dielectric constant. We are interested in this time dependence since it contains information about molecular motion. We can utilize the approach used to obtain equation (8-8), since a capacitor containing a dielectric itself behaves like an RC circuit. This means that the same differential equations apply to the real capacitor as to a fictitious RC circuit that may be treated as its analogue. One of the simplest of such circuits is that represented in Figure 8-1 and described by equation (8-8).

In order to proceed, it is necessary to introduce several definitions from elementary electrostatic theory.

For an infinite-plate capacitor with parallel plates:

$$E = \frac{V}{d}$$ (8-10)

where E is the electric field and d is the distance between the plates. The charge density σ may be defined as

$$\sigma = \frac{Q}{A}$$ (8-11)

where A is the cross-sectional area of the plates. With a vacuum between the plates, the charge density is σ_0 and[†]

$$E = 4\pi\sigma_0$$ (8-12)

[†]The relationship between the electric field strength and the charge density in our capacitor derives from the application of Gauss's flux theorem, which is beyond the scope of this book. Standard texts on electricity and magnetism, such as that by A. D. Kip (McGraw-Hill, New York, 1969), treat this subject in depth. The student should be forewarned, however, that numerous subtle differences in notation and definition abound in this area. The same symbol is often used for different, although related, quantities in various texts.

Now suppose that while the capacitor was connected to a constant voltage source, a dielectric (such as the sample under study) was inserted in the space between the capacitor plates. Additional current would flow into the capacitor owing to polarization of the dielectric material. The polarization itself is time dependent but, at equilibrium, the original charge density σ_0 would have increased to σ, where

$$\sigma = \varepsilon_R \sigma_0 \tag{8-13}$$

ε_R is the limiting value of the dielectric constant of the sample at long times.
   The polarization at infinite time, P_R, is defined by

$$\sigma_0 + P_R = \sigma \tag{8-14}$$

and represents an increase in the charge density due to the presence of the dielectric. The physical origin of the polarization, although not of immediate interest in this discussion of phenomenology, is the displacement of positive and negative charges within the dielectric as well as the reorienting of permanent molecular dipoles under the influence of the electric field. Nevertheless, the polarization may be regarded as resulting from the reorientation of the "macroscopic dipole moment" of the sample under investigation.
   By combining equations (8-12, 8-13, and 8-14), we obtain an expression for the limiting value of the electric polarization at long times in terms of ε_R:

$$\varepsilon_R - 1 = \frac{4\pi}{E} P_R \tag{8-15}$$

Although the polarization is generally time dependent, experimentally one finds that it can be partitioned into an instantaneous component, which we will designate P_u, and a time-dependent term $P_D(t)$. Thus

$$P(t) = P_D(t) + P_u \tag{8-16}$$

at infinite time (equilibrium)

$$P_D(t) = P_D \tag{8-17}$$

and

$$P_R = P_D + P_u \tag{8-18}$$

Thus, equation (8-15) becomes

$$\varepsilon_R - 1 = \frac{4\pi}{E}(P_D + P_u) \tag{8-19}$$

Without loss of generality, the limiting value of the dielectric constant may also be partitioned in the same way, that is, $\varepsilon_R = \varepsilon_D + \varepsilon_u$. By analogy with equation (8-15), we define ε_u as

$$\varepsilon_u - 1 = \frac{4\pi}{E}P_u \tag{8-20}$$

We may now specifically consider the kinetics of polarization. We may assume that the polarization, $P(t)$, approaches its equilibrium value, P_R, at a rate proportional to its distance from equilibrium [see equation (4-23), which is closely related]:

$$\frac{dP(t)}{dt} = -\frac{P(t) - P_R}{\tau} \tag{8-21}$$

This τ is not the same as that in equation (8-7). Via equations (8-16 and 8-18), this simplifies to

$$\frac{dP_D(t)}{dt} = -\frac{P_D(t) - P_D}{\tau} \tag{8-22}$$

Eliminating P_D through equations 8-18, 8-15, and 8-20 gives

$$\tau \frac{dP_D(t)}{dt} + P_D(t) = (\varepsilon_R - \varepsilon_u)\frac{E(t)}{4\pi} \tag{8-23}$$

where the electric field is time dependent $[E = E(t)]$. Now if the real capacitor, which we are viewing as a fictitious RC circuit, is subjected to an electric field periodic in time (see equation 2-53),

$$E^*(\omega, t) = E_0 e^{i\omega t} \tag{8-24}$$

we may solve equation (8-23) to yield

$$P_D^*(\omega) = \frac{\varepsilon_R - \varepsilon_u}{1 + i\omega\tau}\frac{E^*(\omega)}{4\pi} \tag{8-25}$$

The integration of equation (8-23) is carried out in the same fashion as equation (8-8). In addition, it has been assumed that the oscillatory field has been applied to the sample long enough ($t \gg \tau$) so that a steady condition has developed. This being the case, terms involving $e^{-t/\tau}$ are dropped since this factor approaches zero. These exponentials are, in fact, associated with initial transients, arising at the start of the experiment, which quickly become unimportant under most experimental conditions. The complex dielectric constant as a function of frequency can now be defined as

$$\varepsilon^*(\omega) = \frac{4\pi P_D^*(\omega, t)}{E^*(\omega, t)} + \varepsilon_u \qquad (8\text{-}26)$$

where the ε_u term on the right must be included since $P_D^*(\omega, t)$ contains no information about time-independent (instantaneous) properties of the dielectric. Combining equations 8-25 and 8-26 gives

$$\varepsilon^*(\omega) - \varepsilon_u = \frac{\varepsilon_R - \varepsilon_u}{1 + i\omega\tau} \qquad (8\text{-}27)$$

Separating $\varepsilon^*(\omega)$ into its real and imaginary parts in the usual manner (see equation 2-22):

$$\varepsilon^*(\omega) = \varepsilon' - i\varepsilon''$$

$$\varepsilon' = \varepsilon_u + \frac{\varepsilon_R - \varepsilon_u}{1 + \omega^2\tau^2}$$

$$\varepsilon'' = \frac{(\varepsilon_R - \varepsilon_u)\omega\tau}{1 + \omega^2\tau^2} \qquad (8\text{-}28)$$

$$\tan\delta = \frac{\varepsilon''}{\varepsilon'}$$

Equations (8-28) are plotted in Figure 8-2. They are formally identical to the compliance response of a Voigt element in series with a spring when the entire model is subjected to a sinusoidal stress. The complex dielectric constant is thus the analogue of the complex compliance, with the electric field playing the role of stress and the electric displacement $4\pi\sigma$ playing the role of strain.

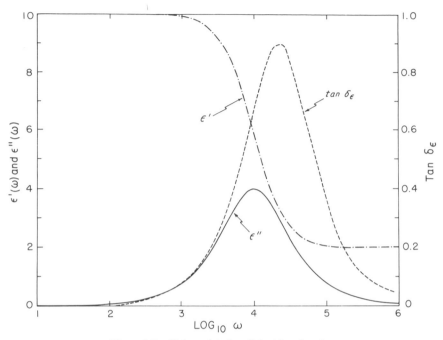

Figure 8-2. Debye plots for dielectric relaxation.

B. MOLECULAR INTEPRETATION OF THE DIELECTRIC CONSTANT

In Section A we developed the concept of polarization under the influence of an electric field from a phenomenological point of view. Now we direct our attention to the underlying molecular properties that give rise to this behavior.

For our purposes, only two molecular mechanisms by which polarization arises need be discussed. First, consider a molecule that has an asymmetrical distribution of positive and negative charges. HCl is an example; the large disparity in the electronegativities of chlorine and hydrogen causes the bonding electron distribution in the molecule to be denser on the halogen than on the hydrogen. This asymmetrical charge distribution in the molecule gives rise to a permanent dipole moment μ which has a value of 1.08 D (D = Debye). For two charges of opposite sign having magnitudes of the charge of the electron, separated by a distance of 1 Å,

$$\mu \equiv er = (4.8 \times 10^{-10} \text{ esu})(1 \times 10^{-8} \text{ cm})$$

$$= 4.8 \times 10^{-18} \text{ esu cm} = 4.8 \text{ D} \tag{8-29}$$

When such a permanent dipole is placed in an electric field, orientation takes place as the molecule attempts to align with the field in order to adopt a low energy configuration. This orientation is clearly time dependent and gives rise to $P_D(t)$, the time-dependent macroscopic polarization discussed previously.

In addition to orienting dipoles, electric fields *induce* dipole moments in molecules, since electrons and nuclei experience forces in opposite directions in the same electric field and since electrons, being less massive, move much more easily than nuclei in a field. The quantity that measures the ease with which the electron cloud in a certain molecule can be distorted is the molecular polarizability α_0. The magnitude of an induced dipole is given as

$$\mu = \alpha_0 E \tag{8-30}$$

Since the electrons move very rapidly because of their small mass, the electronic distortion produced by an electric field is essentially instantaneous. Thus the formation of induced dipoles gives rise to the time-independent P_u term described above.

It remains now to relate the molecular quantities α_0 and μ to the macroscopic polarizability or dielectric constant, which can be measured experimentally. This is a very difficult task and will not be carried out in a rigorous fashion here. Rather, we start our discussion with an approximate equation, given by Debye, which describes the complex dielectric constant in terms of molecular properties. We rationalize the form of the equation through the Clausius–Mosotti equation and then show how $\varepsilon'(\omega)$ and $\varepsilon''(\omega)$ can be derived from this expression. Additional factors that were not included in Debye's original work, such as the effect of the reaction field and orientation correlation—which are important in condensed phases—will also be discussed before extending the treatment to dielectric relaxation in polymers.

Debye showed[2] that

$$\frac{\varepsilon^* - 1}{\varepsilon^* + 2} \frac{M}{d} = \frac{4\pi N_A}{3}\alpha_0 + \frac{4\pi N_A \mu^2}{9kT}\frac{1}{1 + i\omega\tau} \tag{8-31}$$

This equation is derived for a pure substance with molecular weight M and density d. Each molecule of the substance has a permanent dipole moment μ and polarizability α_0. $N_A =$ Avogadro's number.

Our explanation of equation (8-31) makes use of the Clausius–Mosotti equation, which is derived as follows:

The electric field E', produced inside a sphere of uniform dielectric when placed into an electric field E, is given as

$$E' = \frac{3}{\varepsilon + 2} E \qquad (8\text{-}32)$$

This internal field is less than the external field because of polarization of the dielectric in the external field. Clearly, when $\varepsilon = 1$, that is, for a vacuum, $E' = E$.

The total electric moment, M_s, induced in the sphere is just

$$M_s = \alpha_s E \qquad (8\text{-}33)$$

According to equation (8-20), however, the instantaneous polarization P_u is related to the corresponding dielectric constant as

$$P_u = \frac{\varepsilon_u - 1}{4\pi} E' \qquad (8\text{-}34)$$

Since the polarization is the electric moment per unit volume,[†]

$$M_s = P_u V = P_u \tfrac{4}{3}\pi a_s^3 \qquad (8\text{-}35)$$

where a_s is the radius of the dielectric sphere. Combining equations (8-32) through (8-35) leads to

$$\frac{\varepsilon_u - 1}{\varepsilon_u + 2} = \frac{\alpha_s}{a_s^3} \qquad (8\text{-}36)$$

a macroscopic form of the Clausius–Mosotti equation.

To extend this relationship to the molecular domain, suppose that the dielectric sphere contains N_s molecules, each of which has polarizability α_0. Then,

$$\alpha_s = N_s \alpha_0 \qquad (8\text{-}37)$$

Furthermore, assuming that the molecules are themselves spheres with radii a gives

$$\tfrac{4}{3}\pi a^3 = \frac{V}{N_s} = \frac{4\pi a_s^3}{3 N_s} \qquad (8\text{-}38)$$

[†] As defined in equation (8-15), polarization has the same units as σ, charge density. However, in developing the definition, we employed a parallel plate capacitor with plate spacing d. Multiplication of P by d/d shows that polarization may be regarded as electric moment (Q_d) per unit volume (A_d).

so that equation (8-36) becomes

$$\frac{\varepsilon_u - 1}{\varepsilon_u + 2} = \frac{N_s \alpha_0}{N_s a^3} = \frac{\alpha_0}{a^3} \tag{8-39}$$

Furthermore, since the molecular volume can be written as $\frac{4}{3}\pi a^3$ or $M/(dN_A)$,

$$\frac{M}{d}\frac{\varepsilon_u - 1}{\varepsilon_u + 2} = \alpha_0 \frac{4\pi}{3} N_A \tag{8-40}$$

which is the usual form of the Clausius–Mosotti equation for a pure material. It relates the instantaneous value of the dielectric constant ε_u, a macroscopic quantity, to α_0, the molecular polarizability.

We will now shift our attention from ε_u to ε_R, the long-time limiting value of the dielectric constant. Here the tendency of the permanent dipoles to orient in the electric field becomes important.

The energy, U, of a permanent dipole aligned with an electric field of strength E is

$$U = -E\mu \tag{8-41}$$

If the dipole is not perfectly aligned with the field, but rather is directed at some angle θ to the field direction, this energy becomes

$$U = -E\mu \cos\theta \tag{8-42}$$

In a real material, the tendancy of dipoles to align under the influence of the field is counteracted by molecular collisions, which disrupt order. The Boltzmann equation of statistical mechanics provides a simple method by which the average dipole moment $\bar{\mu}$ in the direction of the electric field can be evaluated in this situation:

$$\bar{\mu} = \frac{\int_0^\pi A e^{(E\mu \cos\theta)/kT}\mu \cos\theta \, 2\pi \sin\theta \, d\theta}{\int_0^\pi A e^{(E\mu \cos\theta)/kT} 2\pi \sin\theta \, d\theta} \tag{8-43}$$

where A is a normalization constant and the term $2\pi \sin\theta \, d\theta$ measures the geometric probability that a dipole has an orientation angle θ in the limit that $E \to 0$. (See Figure 8-3). Since it turns out that experimentally we are interested in conditions where kT is large compared with orienting energies, the exponential in equation (8-44) may be expanded and terms higher than

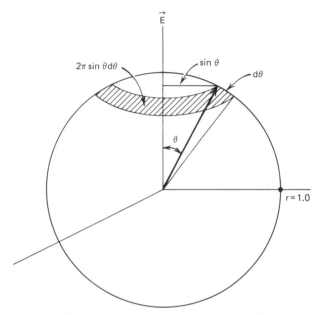

Figure 8-3. Alignment of dipoles in a field.

second order ignored.

$$\bar{\mu} = \frac{\int_0^\pi (1 + E\mu \cos\theta / kT)\mu \cos\theta \sin\theta \, d\theta}{\int_0^\pi (1 + E\mu \cos\theta / kT)\sin\theta \, d\theta} \tag{8-44}$$

This expression is integrable by standard techniques to yield

$$\bar{\mu} = \frac{\mu^2 E}{3kT} \tag{8-45}$$

Debye added this contribution to the induced dipole moment [equation (8-30)] to get

$$\alpha_0 E + \frac{\mu^2 E}{3kT} = \left(\alpha_0 + \frac{\mu^2}{3kT}\right)E$$

or

$$\alpha = \alpha_0 + \frac{\mu^2}{3kT} \tag{8-46}$$

where α is an effective polarizability which measures the tendency for induced dipoles to be formed as well as the tendency for the field to orient the permanent dipoles in the long time-limit.

With this value of α, the Clausius–Mosotti expression may be written for ε_R as

$$\frac{\varepsilon_R - 1}{\varepsilon_R + 2} \frac{M}{d} = \frac{4\pi N_A}{3}\left(\alpha_0 + \frac{\mu^2}{3kT}\right) \tag{8-47}$$

It is now easy to understand the origin of equation (8-31). One sees that it is of the form of the Clausius–Mosotti equation where the complex dielectric constant rather than ε_u or ε_R values are used. The complex formulation introduces a frequency dependence, which appears in the last term of equation (8-31). One would expect the time-dependent contribution to be related to the difference between instantaneous and long-time behavior and, indeed, this is correct, since the factor multiplying the frequency dependence in equation (8-31) is merely the difference between equations (8-47) and (8-40). In fact, these two expressions may be combined with equation (8-47) to yield

$$\frac{\varepsilon^* - 1}{\varepsilon^* + 2} = \frac{\varepsilon_u - 1}{\varepsilon_u + 2} + \left(\frac{\varepsilon_R - 1}{\varepsilon_R + 2} - \frac{\varepsilon_u - 1}{\varepsilon_u + 2}\right)\frac{1}{1 + i\omega\tau} \tag{8-48}$$

Solving for ε^* gives

$$\varepsilon^* = \frac{3\varepsilon_u + 2\left[(\varepsilon_R - 1)Y - \varepsilon_u + 1\right]\left[1/(1 + i\omega\tau)\right]}{3 - \left[(\varepsilon_R - 1)Y - \varepsilon_u + 1\right]\left[1/(1 + i\omega\tau)\right]} \tag{8-49}$$

where

$$Y = \frac{\varepsilon_u + 2}{\varepsilon_R + 2}$$

Clearing fractions, one has

$$\varepsilon^* = \frac{\varepsilon_R + i\varepsilon_u X}{1 + iX} \tag{8-50}$$

where

$$X = \frac{\varepsilon_R + 2}{\varepsilon_u + 2}\omega\tau$$

Multiplication by the complex conjugate of the denominator, collecting real and imaginary terms, and remembering the definition of ε^* in terms of ε' and ε'' in equation (8-28) leads to

$$\varepsilon' = \varepsilon_u + \frac{\varepsilon_R - \varepsilon_u}{1 + X^2}$$

$$\varepsilon'' = \frac{X(\varepsilon_R - \varepsilon_u)}{1 + X^2}$$

(8-51)

It is clear that these equations are closely related to the phenomenological expression equation (8-28) except that the "molecular rotational relaxation time" τ is now replaced by an effective relaxation time τ^* where

$$\tau^* = \frac{X}{\omega} = \frac{\varepsilon_R + 2}{\varepsilon_u + 2} \tau$$

(8-52)

Equations (8-51) and (8-52) are called the Debye equations; taken in conjunction with equation (8-40) and equation (8-47), they relate experimentally measurable macroscopic quantities to molecular properties within the framework and limitations of the model developed by Debye.[2]

Onsager was the first to obtain an expression for the dielectric constant for a more realistic picture of a condensed phase of dipolar molecules.[3] In condensed phases, where molecules are close together, account must be taken of the so-called reaction field. This effect stems from the fact that a dipolar molecule itself polarizes the surrounding medium and this additional polarization reacts back on the molecule. Although Onsager took the reaction field effect into account, he nevertheless neglected orientation correlations between molecules. Derivation of the Onsager equation is beyond the scope of this treatment. The reader is referred to the standard references.[3-5] We merely quote the final result for the dipole moment as a function of the limiting dielectric constants:

$$\mu^2 = \frac{3kT}{4\pi N} \frac{2\varepsilon_R + \varepsilon_u}{3\varepsilon_R} \left(\frac{3}{\varepsilon_u + 2} \right)^2 (\varepsilon_R - \varepsilon_u)$$

(8-53)

where N is the number of molecules per unit volume. It is possible to solve equations similar to equation (8-23) for the Onsager model in an alternating electric field, but the result is quite complex. It can be shown, however, that this treatment gives numerical values close to those obtained from the Debye model.[6]

The orientation correlation between dipoles in condensed phases was considered by Kirkwood and finally evaluated by Fröhlich[7] in terms of the limiting values of the dielectric constant. We shall content ourselves with quoting the Kirkwood–Frölich equation as follows,

$$\varepsilon_R - \varepsilon_u = \frac{3\varepsilon_R}{2\varepsilon_R + \varepsilon_u} \frac{4\pi N}{3kT} \left(\frac{\varepsilon_u + 2}{3} \right)^2 gN\mu^2 \tag{8-54}$$

where g is the Kirkwood correlation factor (see below) and μ is the dipole moment of the isolated molecule.

C. APPLICATION TO POLYMERS

The original forms of the Debye and Onsager treatments are not directly applicable to macromolecules, since they are concerned with assemblies of rigid dipoles in which the magnitudes of the dipole moments do not change as a result of thermal motions. In the case of flexible chain macromolecules whose shapes are constantly changing due to random thermal motions, this is obviously not a realistic approximation.[8] In addition, both the Debye and Onsager treatment give rise to relaxation behavior characterized by a single relaxation time and it is quite clear from the discussion in previous chapters of this book that the single relaxation time model is inadequate to describe the viscoelastic response of polymers.

The Kirkwood–Frölich expression [equation (8-54)] is, however, applicable to flexible chain polymers if we write for the correlation factor

$$g = 1 + \frac{1}{N} \sum_{j=2}^{N} \langle \cos \gamma_{1j} \rangle \tag{8-55}$$

where γ_{1j} is the angle between the first unit of chain (1) and the jth unit and N is the number of repeat units in the chain. μ in equation (8-54) is now understood to refer to the dipole moment of the isolated repeat unit. In the case of the freely jointed chain, when the dipoles of the repeat units lie along the chain contour, it is clear that $g = 1$, since all of the cosine averages are zero (see Chapter 5, Section A). It should also be apparent that g will depend on the chain geometry, and more sophisticated calculations such as those taking into account the interdependence of bond rotational potentials are necessary to obtain values of g for real chains. If the structure of the polymer chain is known exactly, so that the dipole moment of the isolated repeat unit is available together with its angular relationship to the chain backbone, information about the conformational properties of the chain in solution can be obtained by experimental determinations of g through the use of the Kirkwood–Frölich equation. Comparisons may also be made between the experimental value of g and those obtained by theoretical calculations.

Molecular theories describing the dielectric relaxation behavior of polymers have been developed and are summarized in references 5 and 9. Again, if the dipoles are rigidly attached along the chain contour, normal mode theories such as those of Rouse and Bueche described in Chapter 7 for the mechanical case might be expected to be applicable. In addition, the time–temperature superposition principle also generally applies.

A particular treatment of dielectric relaxation data is quite common. This is the so-called Cole–Cole plot[10] obtained by plotting ε'' against ε', each point corresponding to one frequency. From equation (8-28), we have

$$\left(\varepsilon' - \frac{\varepsilon_R + \varepsilon_u}{2}\right)^2 + (\varepsilon'')^2 = \left(\frac{\varepsilon_R - \varepsilon_u}{2}\right)^2 \tag{8-56}$$

which may be verified by direct substitution. Thus, for the single-re-laxation-time model, the Cole–Cole plot is a semicircle of radius $(\varepsilon_R - \varepsilon_u)/2$ with its center on the ε' axis at a distance $(\varepsilon_R - \varepsilon_u)/2$ from the origin. Note that the relaxation time does not appear in equation (8-56) so the Cole–Cole plot is independent of this parameter. An example of a Cole–Cole plot for a single-relaxation-time model with $\varepsilon_R/\varepsilon_u = 5$ is shown in Figure 8-4. In the case of polymers, the single-relaxation-time model is inadequate and Cole–Cole plots are not semicircles. Figure 8-5 is an experimental Cole–Cole plot for the dielectric relaxation of poly (2-chlorostyrene) in the glass transition region. It can be seen that the semicircle of the single-relaxation-time model has become flattened and shows pronounced asymmetry at the

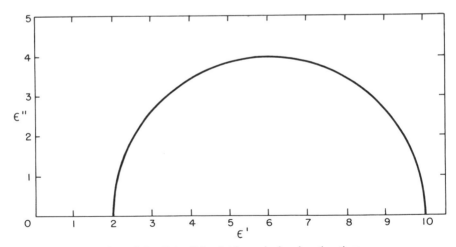

Figure 8-4.   Cole–Cole plot for a single relaxation time.

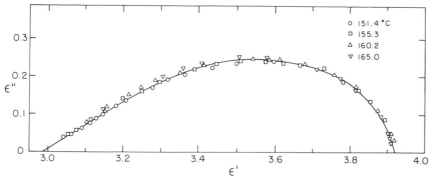

Figure 8-5. Cole–Cole plot for the α relaxation in poly(2 chlorostyrene).

high-frequency end (low values of ε' and ε''). This behavior is typical of that for most amorphous polymers in their primary relaxation regions. A number of empirical distribution functions have been proposed to fit such data. Some of these are described in references 5 and 9. It should be noted that the Cole–Cole method can also be applied to dynamic mechanical data, but this is not often done because most such data are collected over very restricted frequency ranges.

D. EXPERIMENTAL METHODS

Among the attractive features of the dielectric relaxation method are its relative ease of application and the availability of a very large frequency range in a more or less continuous manner. In fact, measurements can be made over the frequency range from 10^{-4} Hz to 3×10^{10} Hz, using a variety of techniques. These are summarized in Table 8-1.

Table 8-1. Methods for Dielectric Measurements

Method	Frequency Range (Hz)
DC transient	10^{-4} to 10^{-1}
Ultra-low-frequency bridge (Harris bridge)	10^{-2} to 10^{2}
Schering bridge; transformer bridge;	
transformer ratio arm bridge	10 to 10^{7}
Resonance circuits; Q meters	10^{5} to 10^{8}
Coaxial (slotted) line; reentrant cavity	10^{8} to 10^{9}
Coaxial line and waveguide	10^{9} to 3×10^{10}

1. DC Transient Current Method

In this method a step voltage is applied to the sample and the current response is measured by a just-response electrometer. For the single-relaxation-time model, the current response would be given by equation (8-9). In recent years this method has been of renewed interest because with the advent of modern computing facilities, it is possible to Fourier-transform the time domain response in order to obtain the frequency response. Several Fourier-transform dielectric spectrometers have been designed. We may note the one due to Johnson et al.[11] The method has the great advantage that a complete dielectric relaxation spectrum can be recorded in a reasonably short time. Practical considerations limit the frequency range available from about 10^{-4} Hz to 10^4 Hz. However, this is a very convenient range for the study of molecular motion in polymers.

2. Bridge Methods

The most commonly employed technique for measuring dielectric relaxation in polymers makes use of the transformer ratio arm bridge. A schematic of such a bridge appears in Figure 8-6. Referring to this figure, the resistance and capacitance of the sample R_s and C_s are balanced by the reference elements R_r and C_r, which have voltages V_R and V_C across them. The conditions for the capacitive and resistive balances are then

$$V_s C_s = V_c C_r$$
$$V_s R_s = V_R R_r \tag{8-57}$$

It is thus possible to balance the bridge by varying the reference resistor R_r and capacitor C_r or by varying the voltages V_R and V_C. Transformer ratio arm bridges may be conveniently employed over a frequency range from

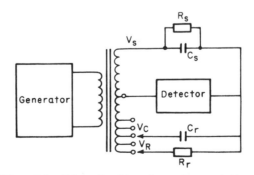

Figure 8-6. Schematic of transform-ratio-arm bridge.

about 10 Hz to 10^6 Hz. Other bridge designs have been described so that the frequency range available by these methods can be said to extend from 10^{-2} Hz to 10^7 Hz.

3.   Other Methods

Bridge methods cannot be used above about 10^7 Hz because the effects of stray inductance become increasingly important at high frequencies. According to Table 8-1, resonance circuits may be used at frequencies up to 10^8 Hz. Above this range, microwave technology must be employed. Since measurements in this frequency range are somewhat more difficult to make and are rarely done on polymers, we shall content ourselves with merely mentioning the existence of appropriate techniques without explaining them further. All the techniques listed in Table 8-1, up through resonance techniques, are referred to as lumped circuit methods. They all have in common that the capacitance of the sample to be measured can be represented by a series model consisting of a capacitance and a resistor, as in Figure 8-1. The lumped circuit methods also have in common that the polymer sample to be examined can be conveniently arranged in the form of a circular disc that is mounted between metal electrodes. The best method of assuring good electrical contact between the polymer and the electrodes is to evaporate a thin film of a conducting metal such as silver onto the polymer surface. This is rarely done in practice, however. It is more common to attach aluminum foil to the sample surface, using silicone grease. Care must be taken to ensure that the sample disc is truly flat and has a smooth surface. Sample cells used to investigate the dielectric relaxation behavior of polymers are usually homemade, although several have been described in the literature.[5,9] It is a relatively easy matter to adapt such cells for the measurement of polymeric liquids as well as solids.

The methods listed in Table 8-1 for the frequency range above 10^8 Hz are referred to as distributed circuit methods. The analysis of the circuits in such methods is not so straightforward as in the case of the lumped circuit methods and the sample cell and geometry are considerably more complex also.

E.   APPLICATION OF DIELECTRIC RELAXATION TO POLYMETHYL METHACRYLATE

In Chapter 4 we cited dynamic mechanical relaxation data for polymethyl methacrylate (PMMA). There it was shown that PMMA possesses two mechanical relaxation regions over the temperature range $-50°$ to $160°$ at

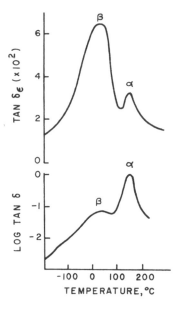

Figure 8-7. Comparison between tan δ_ϵ and tan δ for PMMA.

low frequencies. These were labeled α for the relaxation accompanying the glass transition and β for a secondary relaxation that has generally been associated with motions of the ester side group. PMMA has a predominantly nonpolar backbone with flexible polar side groups. It would therefore be expected that motions involving the side groups are very prominent dielectrically. Figure 8-7 shows a comparison between the dielectric loss tangent and the mechanical loss tangent in PMMA as a function of temperature. It can be seen that the dielectric β relaxation is of much greater magnitude than the mechanical β relaxation, while just the reverse holds true for the dielectric and mechanical α processes. It is generally accepted that the β relaxation in PMMA is due to the hindered rotation of the —$COOCH_3$ group about the carbon–carbon bond linking it to the main chain. The steric hindrance to this rotation comes mainly from the α methyl substituents of the two adjacent repeat units.

The data cited in Figure 8-7 are for "conventional" (or somewhat syndiotactic) PMMA. In the case of isotactic PMMA, the glass transition temperature is reduced so that the α and β relaxations merge even at low frequencies and the dielectric β relaxation appears as a shoulder on the much larger α relaxation. It must be assumed in this case that the onset of side-chain ester group rotation corresponds to the onset of main-chain micro-Brownian motion so that the magnitude of the dielectric α relaxation is enhanced.

F. COMPARISONS BETWEEN MECHANICAL AND DIELECTRIC RELAXATION FOR POLYMERS

It would appear from the foregoing discussion that a correspondence between dielectric and mechanical relaxation can be expected when the molecular motions responsible for a mechanical relaxation involve reorientation of a polar group. Since there is a formal analogy between the complex mechanical compliance J^* and the complex dielectic constant ε^*, it would seem that comparisons between dielectric and mechanical relaxation data is best made by comparing these two quantities or their in-phase and out-of-phase components. This is rarely done in practice, however. It is much more common to construct so-called correlation diagrams in which the frequency of maximum loss is plotted against reciprocal temperature. If the positions of a relaxation region on such a diagram correspond for various relaxational techniques, it can be assumed that the relaxation arises from the same underlying motions. Such a correlation map is shown for the

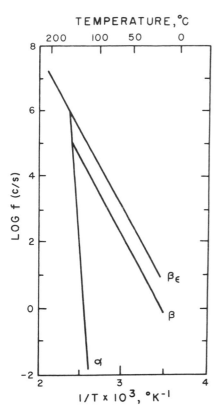

Figure 8-8. Correlation map for PMMA.

α and β relaxations in PMMA in Figure 8-8. It can be seen that the mechanical and dielectric data correspond well for the α relaxation region, but the positions of the mechanical and dielectric β relaxations do not coincide although the lines are parallel. Despite this discrepancy, the two β relaxations are both understood to arise from motions of the ester side-chain, partly because of their similar temperature dependencies. It can also be seen from Figure 8-8 that the α and β relaxations merge at frequencies in excess of about 5×10^5 Hz.

PROBLEMS

1. Consider an electrical circuit consisting of a capacitor and resistor in series and a second capacitor in parallel with the whole array.

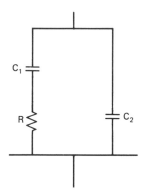

Show that, for a step function in voltage, the capacitance of the circuit is

$$C(t) = (C_R - C_u)(1 - e^{-t/\tau}) + C_u$$

where

$$C_R = C_1 + C_2, \quad \tau = C_1 R, \quad \text{and} \quad C_u = C_2$$

2. Consider a freely jointed chain of N links, each having a dipole moment of magnitude μ_0 along the chain contour.

(a) In the absence of fixed valence angles and bond rotational potentials, calculate the mean square dipole moment of the chain, $\langle \mu^2 \rangle$.

(b) Repeat the calculation in (a) for a chain with tetrahedral (fixed) valence angles. What is the value of g for such a chain?

(c)   If the chain in (b) possesses barriers to rotations about backbone bonds, what would be the qualitative effect on the value of g?

(d)   If the barriers to rotation about the backbone bonds are such that a given bond can only assume rotation angles of $\pi/2$, π, and $3\pi/2$ with respect to the first bond in the chain, and the bond rotation potentials are independent of one another, what is the value of g and $\langle \mu^2 \rangle$?

3.   Derive equation (8-56) from equation (8-28).

4.   It was shown that a "real" capacitor containing a dielectric behaves by itself like an RC circuit. Note the similarity between equations (8-8) and (8-23). If the real capacitor is regarded as a "black box," relate ε' and ε'' to the equivalent *series* capacitance, C_s, and resistance, R_s.

5.   Calculate the power dissipated as a function of ω for a Debye dielectric subjected to a sinusoidal voltage. $\omega = 2\pi f$

6.   Show that ε^* defined by equation (8-28) is formally identical to the compliance, D^*, of a Voigt element in series with a spring.

REFERENCES

1. V. V. Daniel, *Dielectric Relaxation*, Academic Press, New York, 1967.

2. P. Debye, *Polar Molecules*, Dover Publications, New York, 1945.

3. L. Onsager, *J. Am. Chem. Soc.*, **58**, 1486 (1936).

4. C. P. Smyth, *Dielectric Behavior and Structure*, McGraw-Hill, New York, 1955.

5. N. G. McCrum, B. E. Read, and G. Williams, *Anelastic and Dielectric Effects in Polymeric Solids*, Wiley, New York, 1967.

6. N. E. Hill, *Proc. Phys. Soc.*, **78**, 311 (1961).

7. H. Frölich, *Theory of Dielectrics*, 2nd ed., Oxford University Press, Oxford, 1958.

8. P. J. Flory, *Statistical Mechanics of Chain Molecules*, Interscience, New York, 1969.

9. Peter Hedvig, *Dielectric Spectroscopy of Polymer*, Wiley (Halsted), New York, 1971.

10. R. H. Cole and K. S. Cole, *J. Chem. Phys.*, **9**, 341 (1941).

11. G. E. Johnson, E. W. Anderson, G. Z. Link, and D. W. McCall, *Bull. Chem. Phys. Soc.*, **19**(3), 266 (1974).

9
Chemical Stress Relaxation

We have treated the behavior of polymers whose chemical nature remains constant throughout the time of experimentation. In these cases, the time-dependent viscoelastic properties were ascribed to the molecular dynamics inherent in the given macromolecules. Now we turn our attention to a type of time-dependent behavior that results from changes in the actual molecular composition of the sample. Such phenomena are in the area of chemical stress relaxation.

A. DEFINITION OF THE METHOD

The equilibrium stress experienced by a crosslinked rubber network is given by the equation of state for rubber elasticity, developed in Chapter 6 [equation (6-55)]:

$$\sigma = N_0 RT \left[\frac{\overline{r_0^2}}{\overline{r_f^2}} \right] \left(\lambda - \frac{1}{\lambda^2} \right) \qquad (9\text{-}1)$$

A stress relaxation experiment carried out on a crosslinked rubber will thus result in a constant stress, σ, that is proportional to the concentration of network chains. Inasmuch as there is no independent means of estimating the "front factor" $(\overline{r_0^2}/\overline{r_f^2})$, and in view of the other approximations inherent in equation (9-1), we take the front factor to be equal to unity in the ensuing discussion. Thus, we may rewrite equation (9-1) as

$$\sigma = N_0 RT \left(\lambda - \frac{1}{\lambda^2} \right) \tag{9-2}$$

Consider now a crosslinked rubber that is undergoing a chemical reaction resulting in chain scission. If the material is subjected to stress relaxation during this process, it is clear that an equilibrium value of stress will not be attained since the concentration of network chains is not constant. We may rewrite equation (9-2) for this situation as

$$\sigma(t) = N(t) RT \left(\lambda - \frac{1}{\lambda^2} \right) \tag{9-3}$$

where now both the stress and the concentration of network chains are functions of time. In the limit of long times, the final result of the stress relaxation experiment performed on a degrading network will be the decay of stress to zero, just as in the case of a linear amorphous polymer above T_g (Chapter 3, Section A). At the beginning of the stress relaxation experiment, equation (9-3) becomes

$$\sigma(0) = N(0) RT \left(\lambda - \frac{1}{\lambda^2} \right) \tag{9-4}$$

Dividing equation (9-3) by equation (9-4) results in

$$\frac{\sigma(t)}{\sigma(0)} = \frac{N(t)}{N(0)} \tag{9-5}$$

Equation (9-5) provides a direct measure of the extent of degradation of a crosslinked rubber and forms the basis of the chemical stress relaxation technique. That is, measurement of the stress at constant temperature and strain in a degrading rubber network provides a measure of the concentration of network chains as a function of time through equation (9-5).

B. THE TWO-NETWORK THEORY

There are two major limitations to the method presented above. The first of these lies in the use of equation (9-2). It was made clear in the development of Chapter 6 that equation (9-2) is only a first approximation to the true behavior of the system and chemical stress relaxation thus is subject to all the approximations involved in the simple rubber elasticity theory. The second limitation is somewhat less obvious and depends on the details of the degradation process occurring in the network under investigation. The central point is that $N(t)$ in equation (9-5) refers to all the load-bearing network chains. If the degradation process involves additional crosslinking as well as chain scission, it is possible that some of the new network chains formed during degradation may be load-bearing and some may not. If this is the case, equation (9-5) no longer provides a measure of the total concentration of network chains at any time, and some additional assumption must be introduced in order to deal with the situation quantitatively. It is therefore generally assumed that any new network chains formed during degradation are non-load-bearing members of the network and thus do not contribute to $N(t)$ in equation (9-5). This is known as the "two-network" hypothesis; it was first introduced by Andrews, Tobolsky, and Hanson.[1] The validity of the two-network hypothesis has been the subject of several investigations[2,3] and the question of its general applicability remains open. Because of its simplicity and success in interpreting data in many systems, we assume it to be correct in the discussion that follows unless specifically noted otherwise.

The application of the two-network hypothesis to experimental data involves two types of relaxation measurements, continuous and intermittent. The continuous measurement is the "normal" stress relaxation experiment already discussed, in which the polymer is stretched to a fixed length as quickly as possible and maintained at that length and at constant temperature while the stress is followed as a function of time. In such a measurement, $N(t)$ can be calculated from equation (9-5). However, $N(t)$ obtained in this way refers only to the load-bearing chains in the network; in light of the two-network theory, any new chains formed as a result of degradation will contribute nothing to the stress. As a result, $N(t)$ is that portion of the network that has not undergone chain scission at time t. In addition, any chains that have undergone scission at time t and are then reformed cannot contribute to $N(t)$. In the intermittent measurement, the sample is stretched to a fixed length as quickly as possible, the stress is measured, and the sample is returned to its original, unstrained condition. This process is repeated at time intervals such that the sample remains unstrained for much longer periods than it is strained. In such a situation, we may again calculate $N(t)$ from equation (9-5). Now, however, $N(t)$ refers to both the portion of the original network remaining at time t and the new network formed by additional crosslinking during degradation. This means that it is possible to separate the degradative process into its

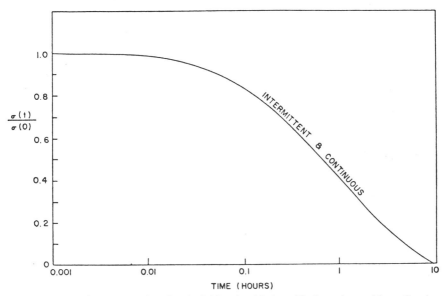

Figure 9-1. $\sigma(t)/\sigma(0)$ versus time for the initiated oxidation of hydrocarbon rubbers. Continuous and intermittent measurements fall on the same curve.

component parts of crosslinking reactions and scission reactions. In the case where degradation involves only chain scission and chains once cut are never reformed, both intermittent and continuous measurements yield identical values for $N(t)$. An example is the initiated oxidation of hydrocarbon rubbers illustrated in Figure 9-1.

In order to interpret the kinetics of the degradation process fully, it is necessary to be able to calculate explicitly the number of scission events occurring in the network as a function of time. Unfortunately, this cannot be done in general without introducing assumptions in addition to those used in arriving at equation (9-5) and the two-network hypothesis. In spite of this, a number of relationships have been derived that have proved very useful in the interpretation of chemical stress relaxation experiments. In all cases, the starting point is equation (9-5). Some of the more important expressions are developed in the following discussion.

C.   IRREVERSIBLE CHAIN SCISSION

The first case to be considered is that in which the degradation process consists of the random scission of network chains. In addition, it is postulated that a chain, once cut, is never re-formed. In such a case of irreversible chain scission, the intermittent and continuous measurements yield identical results, as shown in Figure 9-1, and there is no need to invoke the two-network hypothesis. The most widely applied relationships for computing the time dependence of chain scission under these circumstances are those of Tobolsky et al.[4] and Berry and Watson.[5] The assumptions involved in the Tobolsky equation are that all network chains are elastically active and of the same length, that is, the crosslinks are uniformly distributed. Let q be the number of scission events per unit volume that have occurred at time t; let $M(0)$ be the concentration of chain segments present initially, and let x be the number of segments per chain. (These segments refer to the individual entities in the chain that are capable of undergoing scission. In a hydrocarbon type of rubber the segments are the backbone carbon–carbon links.) The probability that a particular chain segment has been cut at time t must be $q/M(0)$ since q cuts per unit volume have occurred at time t. Also, since a chain segment has either been cut at time t or it has not been cut, the probability that a given segment has not been cut must be $1 - q/M(0)$. Since all the chains have x segments, the fraction of chains per unit volume remaining uncut at time t is $[1 - q/M(0)]^x$. According to the above, we may write

$$N(t) = N(0)\left[1 - \frac{q}{M(0)}\right]^x \tag{9-6}$$

Also

$$N(0) = \frac{M(0)}{x} \tag{9-7}$$

Therefore

$$\frac{N(t)}{N(0)} = \frac{\sigma(t)}{\sigma(0)} = \left[1 - \frac{q}{N(0)x}\right]^x \tag{9-8}$$

For x large, which is always the case within the range of applicability of the kinetic theory of rubber elasticity, the right side of equation (9-8) can be approximated by $e^{-q/N(0)}$ (Problem 9-2).

Substitution into equation (9-8) yields

$$\frac{\sigma(t)}{\sigma(0)} = \exp\left[-\frac{q}{N(0)}\right] \tag{9-9}$$

or

$$q = -N(0)\ln \frac{\sigma(t)}{\sigma(0)} \tag{9-10}$$

Equation (9-10), the Tobolsky equation, has been extensively applied, particularly to the important problem of the oxidative degradation of hydrocarbon networks.

Berry and Watson attempted to refine equation (9-10) by removing the uniform-chain-length assumption on the grounds that it is unrealistic to expect the distribution of crosslinks to be uniform in real networks. To do this, they assumed that the network chains have a range of sizes characterized by the so-called random or "most probable" distribution. This is identical to the distribution of molecular weights that results from most free-radical and condensation–polymerization processes. Familiarity with molecular weight distributions is assumed here. In this case, the concentration of network chains containing x segments becomes

$$N(x) = N(1-p)^{x-1}p \tag{9-11}$$

where N is the total concentration of network chains of all sizes and p is the size parameter (the extent of reaction in the molecular weight distribution case; in the network situation, $p = 1/\bar{x}$).

The initial conditions become

$$N(x,0) = N(0)(1-p)^{x-1}p \tag{9-12}$$

In analogy with the derivation of equation (9-10), the fraction of uncut chains of size z per unit volume at time t is $[1 - q/M(0)]^x$.

Thus

$$N(x,t) = N(x,0)\left[1 - \frac{q}{M(0)}\right]^x \tag{9-13}$$

Substituting equation (9-12) into equation (9-13) results in

$$\frac{N(x,t)}{N(0)} = p(1-p)^{x-1}\left[1 - \frac{q}{M(0)}\right]^x \tag{9-14}$$

In addition, it is clear that

$$\sum_{x=1}^{\infty} N(x,t) = N(t) \tag{9-15}$$

Therefore

$$\sum_{x=1}^{\infty} \frac{N(x,t)}{N(0)} = \frac{N(t)}{N(0)} = \frac{\sigma(t)}{\sigma(0)}$$

$$= \sum_{x=1}^{\infty} p(1-p)^{x-1} \left[1 - \frac{q}{M(0)} \right]^x \qquad (9\text{-}16)$$

The right side of equation (9-16) is a geometric progression that sums to

$$\sum_{x=1}^{\infty} p(1-p)^{x-1} \left[1 - \frac{q}{M(0)} \right]^x = \frac{p[1-q/M(0)]}{1-(1-p)[1-q/M(0)]} \qquad (9\text{-}17)$$

Now, since $M(0) = N(0)/p$, we have

$$\frac{p[1-q/M(0)]}{1-(1-p)[1-q/M(0)]} = \frac{pN(0)-qp^2}{pN(0)-qp^2+qp} \qquad (9\text{-}18)$$

Since $p < 1$, terms in p^2 may be neglected in comparison with terms in p, leading to the result

$$\frac{\sigma(t)}{\sigma(0)} = \frac{N(0)}{N(0)+q} \qquad (9\text{-}19)$$

or

$$q = N(0) \left[\frac{\sigma(0)}{\sigma(t)} - 1 \right] \qquad (9\text{-}20)$$

Equation (9-20) suffers from the same limitations as equation (9-10) except that the restriction of uniform chain length has been removed. The question arises whether the random distribution of chain sizes in a rubber network is more reasonable than uniform chain sizes. The random distribution emphasizes very short chains, and, because of steric requirements, it would not be expected that a real network contains such a preponderance of very short chains. In addition, such short chains, if present, would not obey the Gaussian statistics on which both equation (9-10) and equation (9-20) are fundamentally based. Considerations of this sort lead to the conclusion that equations (9-10) and (9-20) are limiting cases and that the true situation must lie somewhere between these two extremes. From the

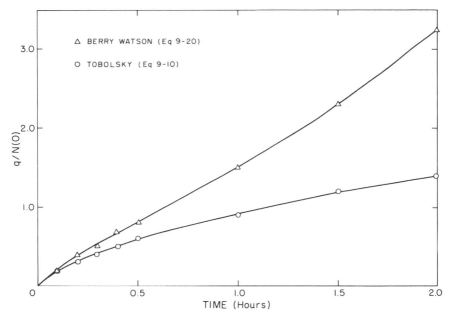

Figure 9-2. Comparison of the time dependence of q as calculated from equation (9-10) and equation (9-20).

practical point of view, a decision must be made as to which relationship should be applied to the system under consideration. It can easily be shown (Problem 9-3 at the end of this chapter) that both equation (9-10) and equation (9-20) give essentially the same results for q in the early stages of the scission process, so that the choice becomes largely a matter of personal preference provided measurements are restricted to the initial phase of the degradation reaction. Figure 9-2 is a plot of q versus time for both equations (9-10) and (9-20) in which the above point is illustrated.

D. REVERSIBLE CHAIN SCISSION

Having treated the situation in which chains once cut are never reformed, we now must establish relationships for reversible scissions in which the cut chains regenerate.[6,7] It will be remembered that in the simplest case of irreversible scission treated already, it is unnecessary to invoke the two-network hypothesis. Any variations of irreversible degradation such as additional crosslinking reactions are still interpretable on the basis of equation (9-10) or (9-20) with the application of the two-network hypothesis. In order to treat the reversible scission case, the two-network hypothesis is necessary even in the simplest situation. In the continuous experiment, only the first scission that occurs in a particular chain is important. No matter how many times the chain is cut, only the first cut will contribute to stress relaxation. Since the chains regenerate, the total concentration of network chains remains at $N(0)$ throughout the experiment, and the intermittent value of $\sigma(t)/\sigma(0)$ will thus also remain unity. In the development of the reversible scission equations, it is again convenient to treat the extreme case of a uniform network on the one hand and a "random" network on the other.

For the uniform network, the probability that a given chain has never been cut at time t becomes $[1-1/N(0)]^q$. Now, by the two-network hypothesis, only the chains that have never been cut at time t are active in supporting the stress. That is, $N(t)$ is the concentration of network chains that have never been cut at time t.

Therefore

$$N(t) = N(0)\left[1 - \frac{1}{N(0)}\right]^q \qquad (9\text{-}21)$$

and

$$\frac{N(t)}{N(0)} = \frac{\sigma(t)}{\sigma(0)} = \left[1 - \frac{1}{N(0)}\right]^q \qquad (9\text{-}22)$$

Expanding the right side of equation (9-22) using the binomial theorem, results in

$$\frac{\sigma(t)}{\sigma(0)} = 1 - \frac{q}{N(0)} + \cdots = \exp\left[\frac{-q}{N(0)}\right] \qquad (9\text{-}23)$$

or

$$q = -N(0)\ln\left[\frac{\sigma(t)}{\sigma(0)}\right] \qquad (9\text{-}24)$$

The result is identical to equation (9-10) obtained for the irreversible scission of a uniform network.

For reversible scission in a random network, the situation is somewhat more complex. In this case, the number of cuts per unit volume in chains that are x segments in length at time t is given by $[qxN(x,0)/\sum xN(x,0)]$.

It immediately follows from this that the probability that a particular chain x segments in length has never been cut at time t is

$$\left[1 - \frac{1}{N(x,0)}\right]^{qxN(x,0)/\sum xN(x,0)}$$

Following the same arguments used in the uniform network case, the concentration of uncut chains of x segments at time t is

$$N(x,t) = N(x,0)\left[1 - \frac{1}{N(x,0)}\right]^{qxN(x,0)/\sum xN(x,0)} \qquad (9\text{-}25)$$

In light of equations (9-11) and (9-14), equation (9-25) may be modified to

$$\frac{\sum N(x,t)}{N(0)} = \frac{N(t)}{N(0)} = \sum_{x=1}^{\infty} (1-p)^{x-1}p\left[1 - \frac{1}{N(x,0)}\right]^{qxN(x,0)/\sum xN(x,0)} \qquad (9\text{-}26)$$

Proceeding in a fashion analogous to that used in deriving equations (9-10), (9-20), and (9-24), the expression

$$\left[1 - \frac{1}{N(x,0)}\right]^{qxN(x,0)/\sum xN(x,0)}$$

is expanded by the binomial theorem

$$\left[1 - \frac{1}{N(x,0)}\right]^{qxN(x,0)/\sum xN(x,0)} = 1 - \frac{qx}{\sum xN(x,0)}$$

$$+ \cdots \cong e^{-qx/\sum xN(x,0)} \qquad (9\text{-}27)$$

Using equation (9-11) it is clear that

$$\sum xN(x,0) = N(0)p\sum x(1-p)^{x-1} = \frac{N(0)}{p} \qquad (9\text{-}28)$$

Since $p = 1/x$; $N(0)/p = M(0)$ and equation (9-27) may be rewritten as

$$\left[1 - \frac{1}{N(x,0)}\right]^{qxN(x,0)/\sum xN(x,0)} = \exp\left[-\frac{qx}{M(0)}\right] \qquad (9\text{-}29)$$

In addition, equation (9-26) becomes

$$\frac{N(t)}{N(0)} = \frac{\sigma(t)}{\sigma(0)} = \sum_{x=1}^{\infty} (1-p)^{x-1} p e^{-qx/M(0)} \tag{9-30}$$

The summation on the right side of equation (9-30) is easily evaluated, since it is simply a geometric series with the first term being $p \exp[-q/M(0)]$ and the common ratio being

$$(1-p)\exp\left[\frac{-q}{M(0)}\right]$$

Thus

$$\frac{\sigma(t)}{\sigma(0)} = \frac{p \exp[-q/M(0)]}{1-(1-p)\exp[-q/M(0)]} \tag{9-31}$$

which yields, upon rearrangement,

$$\frac{q}{M(0)} = \ln\left[p\left(\frac{\sigma(0)}{\sigma(t)} - 1\right) + 1\right] \tag{9-32}$$

E.   CHEMICAL PERMANENT SET

The central role of the two-network hypothesis in chemical stress relaxation has already been dealt with at some length. An experimental test of this hypothesis is possible through the phenomenon of chemical permanent set. In general, if a sample of a degrading rubber network is strained, it will not return to its original length when released but rather will assume a new equilibrium length intermediate between its original length and its strained length. In the limit of long times, when the stress has decayed to zero, the new equilibrium length is identical to the strained length. This is the phenomenon of chemical permanent set. For convenience, we express it as the percentage of the deformation retained by the sample at any time. That is,

$$\text{p.s.} = \frac{L_s - L_u}{L_x - L_u} \times 100 \tag{9-33}$$

where L_s is the new equilibrium unstrained length of the sample at time t,

L_u is the original unstrained length, and L_x is the strained length. In terms of the two-network hypothesis, the permanent set (p.s.) length may be thought of as that length which represents a balance of two opposing forces: the restoring force exerted by the network in equilibrium with the original length of the sample, and the extending force exerted by the network in equilibrium with the strained length of the sample. Using an alternate form of the equation of state for rubber elasticity derived in the Appendix at the end of this chapter, we have

$$\sigma'_u = N_u kT \left[\left(\frac{L_s}{L_u} \right)^2 - \left(\frac{L_u}{L_s} \right) \right] \tag{9-34}$$

$$\sigma'_x = N_x kT \left[\left(\frac{L_s}{L_x} \right)^2 - \left(\frac{L_x}{L_s} \right) \right] \tag{9-35}$$

where σ'_u is the true stress due to the concentration of network chains N_u that are at equilibrium at the original length of the sample, and σ'_x is the true stress due to the concentration of network chains N_x that are in equilibrium with the strained length of the sample. At the permanent set length of the sample:

$$\sigma'_u = - \sigma_x \tag{9-36}$$

From equation (9-34) and equation (9-35):

$$\frac{N_u}{N_x} \left[\left(\frac{L_s}{L_u} \right)^2 - \frac{L_u}{L_s} \right] = \frac{L_x}{L_s} - \left(\frac{L_s}{L_x} \right)^2 \tag{9-37}$$

Substitution of equation (9-33) into equation (9-37) yields an expression for the permanent set on the basis of the two-network theory:

$$\text{p.s.} = C_3 \left[\left(\frac{C_1}{C_2(N_u/N_x)+1} + 1 \right)^{1/3} - 1 \right] \tag{9-38}$$

where $C_1 = (L_x/L_u)^3 - 1$, $C_2 = (L_x/L_u)^2$, $C_3 = 100/[L_x/L_u) - 1]$.

All the quantities in equation (9-38) are experimentally accessible. N_u/N_x is obtainable from the expression

$$\frac{N_u}{N_x} = \frac{\sigma'(t)/\sigma'(0)_{\text{continuous}}}{\sigma'(t)/\sigma'(0)_{\text{intermittent}} - \sigma'(t)/\sigma'(0)_{\text{continuous}}} \tag{9-39}$$

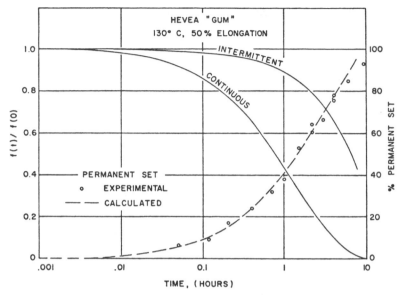

Figure 9-3. Comparison between calculated and observed chemical permanent set for oxidizing hydrocarbon rubber networks. After R. D. Andrews et al., *J. Appl. Phys.*, **17**, 352 (1946), by permission of the authors and the American Institute of Physics.

Equation (9-39) follows from the fact that the continuous measurement is a measure of the concentration of network chains in the sample that have not undergone scission at time t, whereas the intermittent measurement reflects the total concentration of network chains in the sample, including new chains formed as a result of degradation. It is thus possible to calculate the chemical permanent set on the basis of the two-network theory using equations (9-38) and (9-39) and to compare it with the experimental results. Figure 9-3 makes this comparison for oxidizing hydrocarbon rubber networks, and the agreement is seen to be excellent. It is obviously very useful to carry out such a test of the two-network hypothesis when undertaking chemical stress relaxation studies of novel systems.

The necessary machinery has now been developed to examine applications of chemical stress relaxation to specific degrading systems. It is beyond the scope of this treatment to present a detailed account of the chemical stress relaxation method to all the systems that have been studied. Rather, a brief summary is presented in what follows, and the reader is referred to the original literature for more detailed information. Chemical stress relaxation has been used to study networks degrading under the influence of corrosive gases, heat, and radiation of various kinds. Both reversible and irreversible scission processes have been examined.

F.   OXIDATIVE DEGRADATION

An outstanding example of an irreversible scission process is the oxidative degradation of hydrocarbon rubber networks. This process is extremely complex and has been studied by many workers using a variety of techniques. Aside from the intrinsic interest of the problem, its obvious technological importance has led to numerous and extensive investigations. Some of the results of chemical stress relaxation studies on the initiated oxidation of hydrocarbon elastomers will be summarized. (The problem of uninitiated oxidation, or autoxidation, is very much more complex and will not be discussed here. For a review of autoxidation studies, see reference 8.)

The general kinetic scheme for the initiated oxidation of hydrocarbon rubbers is:

Initiation: Initiator $\xrightarrow{k_d}$ 2R'·

$R' \cdot + RH \xrightarrow{k_1} R \cdot + R'H$

Propagation: $R \cdot + O_2 \xrightarrow{k_2} RO_2 \cdot$

$RO_2 \cdot + RH \xrightarrow{k_3} ROOH + R \cdot$

Termination: $2R \cdot \xrightarrow{k_4} R - R$ (recombination)

$RO_2 \cdot + R \cdot \xrightarrow{k_5} ROOR$

$2RO_2 \cdot \xrightarrow{k_6}$ nonradical products

In the above scheme, $R' \cdot$ is a free radical formed by the decomposition of the initiator and $R \cdot$ is a free radical polymer chain.

From the scheme outlined above, it is not possible to identify the kinetic step in which chain scission takes place. However, it is precisely the scission reaction that leads to chemical stress relaxation. Thus, the method provides a convenient means of determining the kinetic step during which the scission reaction takes place. As an illustration, the work of Norling, Lee, and Tobolsky[9] may be cited. These authors studied the benzoyl-peroxide-initiated oxidation of elastomers containing carbon–carbon crosslinks. The Tobolsky equation, equation (9-10), was used to calculate q. From a plot of q as a function of time, the initial rate of scission was determined. Such measurements were carried out with a series of initiator concentrations. The results showed that the initial rate of scission has a linear dependence on initiator concentration (Figure 9-4). Stated mathematically:

$$\left(\frac{dq}{dt}\right)_0 = e_s(2k_d I_0) + \left(\frac{dq}{dt}\right)_{th} \tag{9-40}$$

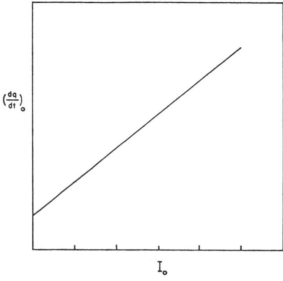

Figure 9-4. Dependence of initial rate of scission on initiator concentration for initiated oxidation of hydrocarbon rubber networks.

where $(dq/dt)_0$ is the initial rate of scission, e_s is the efficiency of scission and reflects the fact that not every radical formed by the decomposition of the benzoyl peroxide is effective in starting oxidation, k_d is the rate constant for the decomposition of the initiator, I_0 is the initiator concentration, and $(dq/dt)_{\text{th}}$ is the rate of scission in the absence of the initiator. k_d is known for benzoyl peroxide from other kinetic investigations so that the slope of a plot of $(dq/dt)_0$ versus I_0 yields values of e_s. Values of e_s for some elastomers obtained in this way are presented in Table 9-1. Intermittent relaxation measurements on these substances indicated that no additional crosslinking took place during the degradation process. This result, coupled with the fact that the scission process is irreversible, showed that there was no need to invoke the two-network hypothesis in interpreting the data. It was also found that e_s is independent of kinetic chain length. This means that scission does not occur in the propagation step, for if scission did take place during propagation, it is apparent that e_s should be a function of the kinetic chain length. It remains to assign the scission process to either the initiation or termination step. In order to do this, further measurements are necessary since the rate of initiation is the same as the rate of termination at the steady state. Since oxygen is not involved in the initiation step, the rate of chemical stress relaxation should not be affected by the absence of

**Table 9-1.   Efficiency of Scission Values,
Benzoyl Peroxide Initiated 60–100°C
(After Norling, Lee, and Tobolsky[9])**

Polymer	e_s
Ethylene–propylene copolymer	0.65
Ethylene–propylene terpolymer	0.60
Amorphous polypropylene	0.60
Polyvinylmethyl ether	0.4
Polypropylene oxide	0.65
Styrene–butadiene coploymer	0.65
Natural rubber	0.6
Polymethylacrylate	0.2
Polyethylacrylate	0.16
Polypropylacrylate	0.07
Polybutylacrylate	0.03

oxygen if scission takes place in the initiation step. Relaxation measurements carried out in a nitrogen atmosphere revealed that the relaxation rate was very much reduced, leading to the conclusion that scission must occur in the termination step. The highest values of e_s in Table 9-1 are 0.60–0.65, which are approximately those for the initiation efficiency of benzoyl peroxide when it is used to initiate free radical polymerizations. Inspection of Table 9-1 shows that the e_s's of the poly-n-alkyl acrylates are anomalously low. This is apparently associated with side-chain oxidation, which would not, of course, contribute to the relaxation process. The efficiencies are seen to decrease with increasing side-chain length, as is to be expected on the basis of this mechanism.

G.   INTERCHANGE NETWORKS: THE POLYSULFIDES

As an example of chemical stress relaxation involving reversible scission may be cited in the case of the polysulfides, or "Thiokol" elastomers.[10] These polymers have the general structure

$$- RS_xRS_xRS_x -$$
$$|$$
$$S_x$$
$$|$$
$$- S_xRS_xRS_xR -$$

where R is an alkyl group and S_x is a polysulfide group. Usually x varies from one to five; it is referred to as the sulfur rank of the polymer. The preparation of the polysulfide elastomers is usually done by reacting an alkyl dihalide with an aqueous solution of sodium polysulfide. Crosslinking is introduced by adding the desired quantity of trihalide to the reaction mixture. As a result of this procedure it is possible to vary x, the nature of R, and the degree of crosslinking. It should be pointed out that aqueous sodium polysulfide solutions always contain a distribution of sizes of polysulfide anions so that it is never possible to obtain a polymer with uniform polysulfide repeat units by this technique. The quoted sulfur rank of the polymer is merely an average value.

The outstanding chemical property of the polysulfide elastomers is their ability to undergo interchange reactions by the scission and recombination of a sulfur–sulfur bond.

$$RS_4R + R'S_4R' \rightleftharpoons 2RS_4R'$$

$$RS_2R + R'S_2R' \overset{\text{cat.}}{\rightleftharpoons} 2RS_2R'$$

These reactions are responsible for the chemical stress relaxation process in the polysulfides. It may be seen that the scission is reversible; this is confirmed experimentally by a study of the intermittent relaxation which remains constant at a value of $\sigma(t)/\sigma(0) = 1$ (Figure 9-5). Chemical stress relaxation has proven useful in the elucidation of the kinetics of scission in these polymers.[10] The experimental results of such studies may be summarized as follows. First, for a polyethylene tetrasulfide series of varying degrees of crosslinking, the relative stress-versus-time curves show simple exponential behavior at all temperatures and all degrees of crosslinking. Stated mathematically:

$$\frac{\sigma(t)}{\sigma(0)} = e^{-k't} \tag{9-41}$$

From plots of $\ln \sigma(t)/\sigma(0)$ versus time at different temperatures, it is possible to obtain k' as a function of temperature. The activation energy for the stress decay process results from a plot of $\log k'$ versus the reciprocal of the absolute temperature. The results of this analysis show that the activation energy remains constant at a value of 25–26 kcal mole^{-1} through a crosslinking range from 10 to 50 mole percent. Having defined the effect of crosslinking, the next variable to be investigated is the sulfur rank. Over a range of sulfur rank from 1.5 to 4.0, it was found that the stress decay is

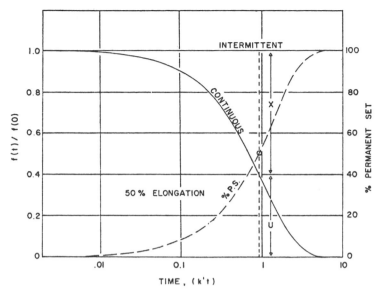

Figure 9-5. Intermittent relaxation versus time for a polysulfide rubber network. After A. V. Tobolsky, *Properties and Structure of Polymers*, Figure 5-20, by permission of John Wiley & Sons, Inc.

well represented by equation (9-41), but that the value of k' increases with increasing sulfur rank; that is, the rate of stress relaxation increases with increasing sulfur rank. The activation energies remain constant at the values of 25–26 kcal mole^{-1} just as in the case of the varying degrees of crosslinking. Finally, a study was conducted on samples containing elemental sulfur physically mixed with the polymers. This was accomplished by immersing the crosslinked samples in molten sulfur, where they were observed to swell and to absorb large quantities of sulfur. When the samples were removed from the sulfur bath, the admixed sulfur showed no tendency to crystallize for long periods of time, remaining in a supercooled liquid condition. The experimental results on these samples again show that the relaxation process obeys equation (9-41) and that the activation energy for stress decay is again 25–26 kcal mole^{-1}. However, the elemental liquid sulfur greatly accelerates the relaxation process in the case of polymers of rank two while leaving the relaxation rate essentially unchanged in the case of polymers of higher rank (rank four). In order to formulate a kinetic scheme for stress decay in the crosslinked polysulfides, the following experimental findings are of significance in addition to the results described above.

1. Eight-membered sulfur rings, long-chain polymeric sulfur, and low-molecular-weight alkyl tetrasulfides all undergo thermal homolytic scission with an activation energy of approximately 35 kcal mole^{-1}.

2. Electron spin resonance and magnetic susceptibility data indicate the existence of free radical chain ends in polymeric sulfur at temperatures of around 300°C.

3. Alkyl disulfides are much more stable thermally than alkyl polysulfides. As noted above, the rate of chemical stress relaxation in the polysulfide polymers increases markedly with increasing sulfur rank.

4. Stress relaxation of polysulfide polymers is accelerated by near ultraviolet light and the activation energy for the process is thereby greatly lowered.

Utilizing all the foregoing experimental findings, the following kinetic scheme is proposed for the chemical stress relaxation of the polysulfides:

Initiation:

$$\sim\!\!\text{S}_4\!\!\sim(\text{strained}) \xrightarrow{k_i} \sim\!\!\text{S}_2\!\cdot\; +\; \cdot\text{S}_2\!\!\sim(\text{unstrained})$$

or, in the case of admixed sulfur,

$$\text{S}_8(\text{ring}) \xrightarrow{k_i} \text{S}_8(\text{diradical chain})$$

Propagation:

$$\sim\!\!\text{S}_2\!\cdot\; +\!\sim\!\!\text{S}_x\!\!\sim(\text{strained}) \xrightarrow{k_p} \sim\!\!\text{S}_x\!\!\sim(\text{unstrained}) +\!\!\sim\!\!\text{S}_2\!\cdot$$

Termination: $\sim\!\!\text{S}_2\!\cdot\; +\; \cdot\text{S}_2\!\!\sim \xrightarrow{k_i} \sim\!\!\text{S}_4\!\!\sim(\text{unstrained})$

All of the reactions written above are, of course, reversible, but the reverse reactions contribute nothing to $N(t)$ by the two-network hypothesis. It remains to establish that the chain reaction presented schematically above, does indeed satisfy the experimental results. The rate of stress decay is given by

$$\frac{-dN(t)}{dt} = k_i[\sim\!\!\text{S}_4\!\!\sim] + k_p[\sim\!\!\text{S}_x\!\!\sim][\sim\!\!\text{S}_2\!\cdot] \qquad (9\text{-}42)$$

The assumption is now made that the observed relaxation takes place mainly in the propagation step. This means that a single scission event in the initiation step produces many scissions in the propagation step. If this is so, the first term of equation (9-42) is small compared to the second and equation (9-42) becomes

$$\frac{-dN(t)}{dt} = k_p[\sim S_x \sim][\sim S_2 \cdot] \tag{9-43}$$

At the stationary state, the rate of initiation is equal to the rate of termination:

$$k_i[\sim S_4 \sim] = k_t[\sim S_2 \cdot]^2 \tag{9-44}$$

or

$$[\sim S_2 \cdot] = \left(\frac{k_i}{k_t}\right)^{1/2}[\sim S_4 \sim]^{1/2} \tag{9-45}$$

and

$$\frac{-dN(t)}{dt} = k_p\left(\frac{k_i}{k_t}\right)^{1/2}[\sim S_4 \sim]^{1/2}[\sim S_x \sim] \tag{9-46}$$

Since it is assumed that stress relaxation occurs in the propagation step, $[\sim S_x \sim]$ can be replaced by $mN(t)$, where m is the number of polysulfide links (linkages susceptible to cleavage per network chain). This is obviously not strictly correct since scission of an S_4 chain in the initiation step also leads to relaxation of stress. However, if the kinetic chain length is great, the error introduced by this assumption will be small. Also, under these circumstances, $[\sim S_4 \sim]$ will remain approximately constant throughout the reaction. Designating $[\sim S_4 \sim]$ by I results in

$$\frac{-dN(t)}{dt} = k_p\left(\frac{k_i}{k_t}\right)^{1/2} I^{1/2} m N(t) \tag{9-47}$$

Note that equation (9-47) is equally valid when the initiator is elemental sulfur. In the case of sulfur as the initiator, it is only necessary to set I in equation (9-47) equal to the concentration of elemental sulfur added to the polymer.

In order to compare equation (9-47) to experiment, it is necessary to return to equation (9-41). By the fundamental relation of chemical stress relaxation, equation (9-41) becomes

$$\frac{\sigma(t)}{\sigma(0)} = \frac{N(t)}{N(0)} = e^{-k't} \tag{9-48}$$

which, upon differentiation, yields

$$\frac{-dN(t)}{dt} = k'N(t) \tag{9-49}$$

Comparing equation (9-49) to equation (9-47),

$$k' = k_p \left(\frac{k_i}{k_t}\right)^{1/2} I^{1/2} m \tag{9-50}$$

At a given temperature, all of the quantities in equation (9-50) are independent of sulfur rank with the exception of I. I increases with sulfur rank since it is only the higher polysulfides that are capable of undergoing thermal homolytic scission. As a result, k' also increases with increasing sulfur rank, leading to an increase in the overall rate of stress relaxation as is actually observed. In the same fashion, if the concentration of added elemental sulfur is increased, k' increases and the rate of stress relaxation at a given temperature again increases. In addition, as the degree of crosslinking increases, the number of polysulfide groups per network chain decreases and m decreases, thus reducing the rate of stress relaxation, again in accord with experiment.

Finally, it is necessary to discuss the temperature dependence of the rate constant. From equation (9-50) it immediately follows that the observed activation energy, H', is given by

$$H' = H_p + \tfrac{1}{2}(H_i - H_t) \tag{9-51}$$

As already stated, H' amounts to 25–26 kcal in all systems studied. In addition, as has been discussed, H_i is 35 kcal mole^{-1}. Since H_t involves a radical recombination reaction, its magnitude is probably quite small and the assumption is thus made that it is equal to zero. On this basis, H_p becomes approximately 8 kcal mole^{-1}. The only direct measurement of the activation energy for the attack of a polysulfide radical on a polysulfide linkage[11] gives an H_p in the range of 3–6 kcal mole^{-1} in fair agreement with the value obtained above.

APPENDIX

In Chapter 6, Section B1, the equation of state for rubber elasticity based
on the statistical theory was shown to be

$$\sigma = N_0 RT \left(\lambda - \frac{1}{\lambda^2} \right) \qquad (9\text{-}2)$$

Here σ is the elastic force per unit *unstrained* cross-sectional area (σ
$= F/A_0$) and is generally called the "nominal stress." This is the definition
of stress used through most of this book. It is possible, however, to define
the stress in terms of the cross-sectional area of the strained rubber:

$$\sigma' = \frac{F}{A} \qquad (a)$$

where σ' is the "true stress."

The equation of state will now be rewritten in terms of the true stress. In
equation (9-2), the concentration of network chains (N_0) is defined as the
number of chains (N) divided by the volume of the unstrained rubber (V_0);
λ is just L/L_0, the ratio of the length of the stretched and unstretched
rubber. Equation (9-2) can be rewritten as

$$\frac{F}{A_0} = \left(\frac{NRT}{V_0} \right) \left(\frac{L}{L_0} - \frac{L_0^2}{L^2} \right)$$

or, since $V_0 = A_0 L_0$,

$$F = \left(\frac{NRT}{L_0} \right) \left(\frac{L}{L_0} - \frac{L_0^2}{L^2} \right) \qquad (b)$$

Now divide equation (b) by the cross-sectional area of the strained sample
to obtain the true stress:

$$\sigma' = \frac{F}{A} = \left(\frac{NRT}{AL_0} \right) \left(\frac{L}{L_0} - \frac{L_0^2}{L^2} \right) \qquad (c)$$

Multiplying by L/L and rearranging, one gets

$$\sigma' = \left(\frac{NRT}{V} \right) \left(\frac{L}{L_0} \right) \left(\frac{L}{L_0} - \frac{L_0^2}{L^2} \right) \qquad (d)$$

However, we know from Chapter 2, Section A, that upon extension the volume of the rubber stays nearly constant; thus,

$$\sigma' = N_0 RT \left(\frac{L^2}{L_0^2} - \frac{L_0}{L} \right) \tag{e}$$

Equation (e) is an alternative expression of the equation of state for rubber elasticity previously given in equation (9-2).

PROBLEMS

1. Plot schematically the time dependence of intermittent and continuous relaxation measurements for degradation processes in which

    (a) Both chain scission and additional crosslinking take place.

    (b) Only chain scission takes place but chains reform after being cut.

2. Show that $[1 - q/N(0)x]^x \approx \exp[-q/N(0)]$ for large x.

3. Show that equation (9-10) and equation (9-20) are equivalent at low values of q.

4. Verify that the leading terms in the binomial expansion of $[1 - 1/N(0)]^q$ are 1 and $-q/N(0)$.

5. Evaluate $N(0)p \sum_{x=1}^{\infty} x(1-p)^{x-1}$.

6. Obtain equation (9-38) from equation (9-33) and equation (9-37) using definitions of C_1, C_2, and C_3.

7. Show that $H' = H_p + \frac{1}{2}(H_i - H_t)$, starting from equation (9-50).

REFERENCES

1. A. V. Tobolsky, *Properties and Structure of Polymers*, Wiley, New York, 1960.

2. P. J. Flory, *Chem. Rev.*, **35**, 51 (1944).

3. D. K. Thomas, *Polymer*, **7**, 125 (1966).

4. A. V. Tobolsky, D. J. Matz, and R. B. Mesrobian, *J. Am. Chem. Soc.*, **72**, 1942 (1950).

5. J. P. Berry and W. F. Watson, *J. Polym. Sci.*, **18**, 204 (1955).

6. H. Yu, *J. Polym. Sci.*, **B2**, 631 (1964).

7. A. V. Tobolsky, *J. Polym. Sci.*, **B2**, 631 (1964).

8. J. R. Dunn and J. Scanlon, "Stress Relaxation Studies of Network Degradation," in *The Chemistry and Physics of Rubberlike Substances*, L. Bateman, Ed., MacLaren, London, 1963.

9. P. M. Norling, T. C. P. Lee, and A. V. Tobolsky, *Rubber Chem. Technol.*, **38**, 1198 (1965).

10. A. V. Tobolsky and W. J. MacKnight, *Polymeric Sulfur and Related Polymers*, Wiley, New York, 1966, Chapter 4.

11. D. K. Gardner and K. Fraenkel, *J. Am. Chem. Soc.*, **78**, 3279 (1956).

Answers to Problems

CHAPTER 2

2-1. The shear strain wanted is given, according to equation (2-5), as

$$\gamma = \frac{\Delta X}{C} = \frac{0.40 \text{ cm}}{2.0 \text{ cm}} = 0.20$$

At $t = 10^{-4}$, the compliance is numerically equal to $10^{-9} \ Pa^{-1}$. Thus, making use of equation (2-9),

$$\sigma_{s_0} = \frac{\gamma(t)}{J(t)} = \frac{\gamma(10^{-4})}{J(10^{-4})} = 0.20 \times 10^9 Pa$$

However, since the sample cross-sectional area, $A \times B$, is 4 cm², that is 4×10^{-4} m², the constant force necessary to observe this strain is

$$F_{s_0} = \sigma_{s_0}(A \times B) = 0.8 \times 10^5 \text{ newtons}$$

and division by the acceleration due to gravity yields

$$m = \frac{F_{s_0}}{a} = 8.2 \times 10^6 \text{ g}$$

Thus 8.2×10^6 g placed on the sample pan in Figure 2-2 would cause the pointer to move down 0.40 cm in 10^{-4} seconds. Clearly all the apparatus is assumed to have an infinite modulus and inertial effects are ignored.

2-2.   Since G' and G'' are given, our problem is to express $\tan \delta$, $|G|^2$, $|J|^2$, J' and J'' in terms of these given parameters. Equation (2-21) gives

$$\tan \delta = \frac{G''}{G'}$$

Next it will be helpful to place the storage and loss components of both the stress and strain on the same set of stress–strain vectors. This is shown in Figure 2-8a, which is merely a composite of Figure 2-7. The various vectors are labeled in the table. For clarity, it is now desirable to rotate the triangle ABF so that side AB of that triangle is coincident with side AC of triangle AEC. This is shown in Figure 2-8b. The vectors designated in the table do not change in magnitude under this transformation.

To proceed, AE represents σ', CE represents σ'', and AC represents σ. If these three quantities are divided by the magnitude of γ, they become G', G'', and $|G|$ respectively. Clearly, then, application of the Pythagorean theorem gives

$$|G|^2 = G'^2 + G''^2$$

Now from Figure 2-8, relationships of congruent triangles give

$$\frac{BF}{AB} = \frac{CE}{AC}$$

In terms of more relevant quantities

$$\frac{\gamma''}{\gamma} = \frac{\sigma''}{\sigma}$$

Arithmetic rearrangement and division by σ gives

$$\frac{\gamma''}{\sigma} = \sigma'' \frac{\gamma}{\sigma^2} \cdot \frac{\gamma}{\gamma} = \frac{\sigma''}{\gamma} \frac{\gamma^2}{\sigma^2}$$

which is equal to

$$J'' = \frac{G''}{|G|^2} = \frac{G''}{G'^2 + G''^2}$$

Using the same technique,

$$J' = \frac{G'}{|G|^2} = \frac{G'}{G'^2 + G''^2}$$

Finally, since $|J|$ and $|G|$ are reciprocals,

$$|J|^2 = \frac{1}{|G|^2} = \frac{1}{G'^2 + G''^2}$$

Parts (b) and (c) are similar.

2-3. Equation (2-28) may be restated as

$$\sigma(t) = \int_{-\infty}^{t} \frac{d\epsilon(\mu)}{d\mu} E(t - \mu) \, d\mu \tag{a}$$

for a tensile experiment. The strain history, in terms of the variable μ, is

$$-\infty < \mu \leqslant 0 \qquad \frac{d\epsilon(\mu)}{d\mu} = 0$$

$$0 < \mu \leqslant t_1 \qquad \frac{d\epsilon(\mu)}{d\mu} = k$$

$$t_1 < \mu \leqslant 2t_1 \qquad \frac{d\epsilon(\mu)}{d\mu} = -k$$

$$2t_1 < \mu \leqslant t \qquad \frac{d\epsilon(\mu)}{d\mu} = 0$$

This strain history can be built into equation (a) to yield

$$\sigma(t) = \int_0^{t_1} k \left[E_1 e^{-[(t-u)/\tau_1]} + E_2 e^{-[(t-u)/\tau_2]} \right] du$$

$$+ \int_{t_1}^{2t_1} (-k) \left[E_1 e^{-[(t-u)/\tau_1]} + E_2 e^{-[(t-u)/\tau_2]} \right] du$$

which can be integrated to give

$$\sigma(t) = k \left\{ E_1 e^{-t/\tau_1} \tau_1 \left[2 e^{t_1/\tau_1} - e^{2t_1/\tau_1} - 1 \right] \right.$$

$$\left. + E_2 e^{-t/\tau_2} \tau_2 \left[2 e^{t_1/\tau_2} - e^{2t_1/\tau_2} - 1 \right] \right\}$$

2-4. Work is usually defined as the integral of $f\,dl$. Here we will consider work per unit volume. Thus

$$W = \int \frac{f\,dl}{V} = \int \frac{f}{A}\frac{dl}{l} = \int \sigma\,d\gamma \tag{a}$$

for a shear experiment. Now let

$$\sigma = \sigma_0 \cos \omega t$$

so that in-phase strain is

$$\gamma = \gamma_0 \cos \omega t$$

Substitution of these functions into equation (a) yields

$$W = \int_0^{2\pi} \sigma_0(\cos \omega t)\gamma_0\,d\cos \omega t$$

which integrates to zero. Thus the work done in going through one cycle of deformation (ωt goes from 0 to 2π) is zero if the stress and strain are in phase.

For out-of-phase strain:

$$\gamma = \gamma_0 \sin \omega t$$

which yields for the integral work

$$W = \int_0^{2\pi} \sigma_0(\cos \omega t)\gamma_0\,d(\sin \omega t)$$

Here integration gives

$$W = \gamma_0 \sigma_0 \pi$$

Clearly work is done when the stress and strain are out of phase.

2-5. Equation (2-49) gives

$$\frac{1}{p^2} = L(G(t))L(J(t)) \tag{2-49}$$

which is a statement of the relationship between $G(t)$ and $J(t)$ in transform space.

The Laplace transform of the $J(t)$ given is

$$L(J(t)) = \frac{A\Gamma(1+m)}{p^{m+1}}$$

which yields upon substitution into equation (2-49) the result for the transform of $G(t)$

$$L(G(t)) = \frac{1}{A\Gamma(m+1)p^{1-m}} \frac{\Gamma(1-m)}{\Gamma(1-m)}$$

the last multiplication term being added to facilitate return from transform space. Thus

$$G(t) = \frac{t^{-m}}{A\Gamma(m+1)\Gamma(1-m)}$$

Rearrangement and use of the expressions given in the problem yield the desired result.

2-6. Equation (2-49) relates $G(t)$ and $J(t)$ in transform space.

$$\frac{1}{p^2} = L(G(t))L(J(t)) \tag{2-49}$$

It is simple to solve in this space and then transform back

$$L(G(t)) = \int_0^\infty e^{-pt} G(t)\, dt$$

$$= \int_0^\infty e^{-pt} G_0 e^{-t/\tau}\, dt = G_0 \int_0^\infty e^{-(p+1/\tau)t}\, dt$$

$$= G_0 \left(p + \frac{1}{\tau} \right)^{-1} \int_0^\infty e^{-x}\, dx = G_0 \left(p + \frac{1}{\tau} \right)^{-1} (-e^{-x}) \Big|_0^\infty$$

$$= G_0 \left(p + \frac{1}{\tau} \right)^{-1}$$

Thus

$$LJ(t) = \frac{1}{p^2} \frac{1}{L(G(\tau))} = \frac{(p+1/\tau)}{p^2 G_0} = \frac{1}{pG_0} + \frac{1}{p^2 G_0 \tau}$$

Transforming back into real space,

$$J(t) = L^{-1}\left(\frac{1}{pG_0}\right) + L^{-1}\frac{1}{p^2 G_0 \tau}$$

$$= \frac{1}{G_0} L^{-1}\left(\frac{1}{p}\right) + \frac{1}{\tau G_0} L^{-1}\left(\frac{1}{p^2}\right)$$

$$= \frac{1}{G_0} + \frac{1}{G_0 \tau} t$$

$$\therefore J(t) = \frac{1}{G_0}\left(1 + \frac{t}{\tau}\right)$$

2-7. From equation (2-56) we know that

$$E'(\omega) = \omega \int_0^\infty \sin \omega s E(s)\, ds$$

$$E''(\omega) = \omega \int_0^\infty \cos \omega s E(s)\, ds$$

Similar expressions also apply to shear, thus for this problem

$$G'(\omega) = \omega \int_0^\infty \sin \omega s G_0 e^{-s/\tau}\, ds$$

$$= G_0 \omega \int_0^\infty (\sin \omega s) e^{-s/\tau}\, ds$$

$$\boxed{\int_0^\infty e^{-ax} \sin mx\, dx = \frac{m}{a^2 + m^2}}$$

$$G'(\omega) = \frac{G_0 \omega^2}{(1/\tau)^2 + \omega^2} = G_0 \frac{\omega^2 \tau^2}{1 + \omega^2 \tau^2}$$

also

$$G''(\omega) = \omega \int_0^\infty \cos \omega s G_0 e^{-s/\tau}\, ds$$

$$= \omega G_0 \int_0^\infty (\cos \omega s) e^{-s/\tau}\, ds$$

$$\boxed{\int_0^\infty e^{-ax} \cos mx\, dx = \frac{a}{a^2 + m^2}}$$

$$G''(\omega) = \frac{\omega G_0 1/\tau}{(1/\tau)^2 + \omega^2} = \frac{G_0 \omega \tau}{1 + \omega^2 \tau^2}$$

2-8.

$$G^* \equiv \frac{\sigma^*}{\gamma^*} = \frac{\sigma_0 e^{i(\omega t + \delta)}}{\gamma_0 e^{i\omega t}} = \frac{\sigma_0 e^{i\delta}}{\gamma_0}$$

but $e^{ix} = \cos x + i \sin x$ so that

$$G^* = \frac{\sigma_0}{\gamma_0}(\cos \delta + i \sin \delta)$$

However, $G^* \equiv G' + iG''$, and we have:

$$G' = \frac{\sigma_0}{\gamma_0} \cos \delta \quad \text{and} \quad G'' = \frac{\sigma_0}{\gamma_0} \sin \delta$$

2-9.

$$\eta^* = \frac{\sigma^*}{d\epsilon^*/dt}$$

For a dynamic experiment, $\epsilon^* = \epsilon_0 e^{i\omega t}$

$$\therefore \frac{d\epsilon^*}{dt} = \epsilon_0 i\omega e^{i\omega t} = i\omega \epsilon^*$$

$$\eta^* = \frac{\sigma^*}{i\omega \epsilon^*}$$

But

$$G^* \equiv \frac{\sigma^*}{\epsilon^*}$$

and

$$\eta^* = \frac{G^*}{i\omega} = \frac{-iG^*}{\omega}$$

$$\eta^* = \eta' - i\eta'' = \frac{-iG^*}{\omega} = \frac{-i}{\omega}(G' + iG'') = \frac{-iG'}{\omega} + \frac{G''}{\omega}$$

Thus

$$\eta' = \frac{G''}{\omega} \quad \text{and} \quad \eta'' = \frac{G'}{\omega}$$

2-10.

$$\gamma(t) = \int_{-\infty}^{t} \frac{d\sigma(u)}{du} J(t-u)\,du \tag{2-27}$$

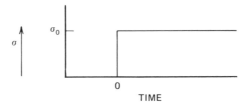

At time zero, the stress is instantaneously increased from 0 to σ_0.
  To observe how the delta function works, consider

$$\sigma(t) = \int_{-\infty}^{t} \frac{d\sigma(u)}{du}\,du$$

as the normal relationship between the stress and stress rate. For an instantaneously applied stress with an infinite stress rate we may write

$$\sigma(t) = \int_{-\infty}^{t} \delta(u)\sigma_0\,du$$

where $\delta(u)$ is the Dirac delta function. According to the definition of this function given in the question

$$\sigma(t) = 0 \qquad t < 0$$

$$\sigma(t) = \sigma_0 \qquad t \geqslant 0$$

Similarly for equation (2-27):

$$\gamma(t) = \int_{-\infty}^{t} \delta(u) \sigma_0 J(t-u) \, du$$

For all $t > 0$ this becomes:

$$\gamma(t) = \sigma_0 J(t)$$

or

$$J(t) = \frac{\gamma(t)}{\sigma_0} \qquad\qquad (2\text{-}9)$$

2-11.

$$\boxed{\frac{d}{da} \int_{p}^{q} f(x,a) \, dx = \int_{p}^{q} \frac{\partial}{\partial a} \big[f(x,a) \big] \, dx + f(q,a) \frac{dq}{da} - f(p,a) \frac{dp}{da}}$$

$$\sigma(t) = \int_{-\infty}^{t} \frac{d\epsilon(u)}{du} E(t-u) \, du$$

$$\frac{d}{dt} \sigma(t) = \frac{d}{dt} \int_{-\infty}^{t} \frac{d\epsilon(u)}{du} E(t-u) \, du$$

$$\dot{\sigma}(t) = \int_{-\infty}^{t} \frac{\partial}{\partial t} \left[\frac{d\epsilon(u)}{du} E(t-u) \right] du + \frac{d\epsilon(t)}{dt} E(0)$$

$$\text{if} \qquad \frac{d\epsilon(t)}{dt} = \begin{cases} k & \text{for } t \geqslant 0 \\ 0 & \text{for } t < 0 \end{cases}$$

$$\dot{\sigma}(t) = \int_{0}^{t} k \frac{\partial}{\partial t} \big[E(t-u) \big] \, du + kE(0)$$

But

$$\frac{\partial}{\partial t} E(t-u) = \frac{dE(t-u)}{d(t-u)} \frac{\partial(t-u)}{\partial t} = \frac{dE(t-u)}{d(t-u)}$$

$$\therefore \dot{\sigma} = \int_0^t k \frac{dE(t-u)}{d(t-u)} \, du + kE(0)$$

or

$$\dot{\sigma} = \int_0^t -kd(E(t-u)) + kE(0) = kE(t-u)\Big|_0^t + kE(0)$$

$$= -kE(0) + kE(t) + kE(0)$$

$$\frac{\dot{\sigma}}{\dot{\gamma}} = E(t)$$

2-12. This will be done in two steps utilizing different forms of the Boltzmann principle.

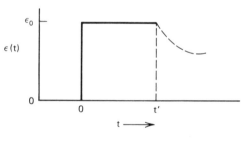

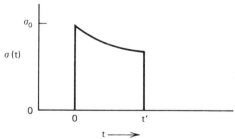

(a)

$$u < 0 \qquad \sigma(u) = 0$$

$$0 \leqslant u \leqslant t' \qquad \sigma(u) = \epsilon_0 E(u)$$

$$t' < u \qquad \sigma(u) = 0$$

Starting off with equation (2-48), an expression for the Boltzmann principle in Laplace space,

$$L(\sigma(t)) = p L(\epsilon(t)) L E(t)$$

$$L(\epsilon(t)) = \frac{1}{p} \frac{L\sigma(t)}{LE(t)}$$

But $LF(t) = \int_0^\infty e^{-pt} F(t)\, dt$ so

$$L\sigma(t) = \int_0^\infty e^{-pt}\sigma(t)\, dt = \int_0^{t'} \epsilon_0 E(t) e^{-pt}\, dt = \int_0^{t'} \epsilon_0 E_0 e^{-t/\tau} e^{-pt}\, dt$$

$$= \epsilon_0 E_0 \int_0^{t'} e^{-(p+1/\tau)t}\, dt = \frac{\epsilon_0 E_0}{p+1/\tau}(-e^{-x})_0^{t'(p+1/\tau)}$$

$$= \frac{E_0 \epsilon_0}{p+1/\tau}\left[1 - e^{-t'(p+1/\tau)}\right]$$

$$LE(t) = \int_0^\infty e^{-pt} E_0 e^{-t/\tau}\, dt = E_0 \int_0^\infty e^{-(p+1/\tau)t}\, dt = \frac{E_0}{p+1/\tau}$$

$$\therefore L\epsilon(t) = \frac{1}{p}\frac{L\sigma(t)}{LE(t)} = \frac{\left[E_0\epsilon_0/(p+1/\tau)\right]\left[1 - e^{-t'(p+1/\tau)}\right]}{pE_0/(p+1/\tau)}$$

$$= \frac{\epsilon_0}{p}\left[1 - e^{-t'(p+1/\tau)}\right]$$

$$= \frac{\epsilon_0}{p} - \frac{\epsilon_0}{p} e^{-t'(p+1/\tau)} = \frac{\epsilon_0}{p} - \frac{\epsilon_0}{p} e^{-t'p} e^{-t'/\tau}$$

Taking inverse transforms,

$$\epsilon(t) = \epsilon_0 - \epsilon_0 e^{-t'/\tau} L^{-1}\left(\frac{e^{-pt'}}{p}\right)$$

From tables,

$$
L^{-1}\frac{e^{-ks}}{s^u} = \begin{cases} 0 & \text{when} \quad 0 < t < k \\ \dfrac{(t-k)^{u-1}}{\Gamma(u)} & \text{when} \quad t > k \end{cases} \qquad \text{for} \quad u > 0
$$

In our case $u = 1$, which is greater than zero. Also $k = t'$ and $t > t'$ always, since we are interested in strains after the stress relaxation experiment. Thus, the general situation holds and

$$
\epsilon(t) = \epsilon_0 - \epsilon_0 e^{-t'/\tau}\left[\frac{(t-t')^0}{\Gamma(1)}\right]
$$

$$
(t-t')^0 = 1 \quad \text{and} \quad \Gamma(1) = 0! \equiv 1
$$

Thus,

$$
\epsilon(t) = \epsilon_0(1 - e^{-t'/\tau})
$$

(b)   The other technique involves transforming $\epsilon(t)$ directly to $D(t)$ via the techniques used in problem 2-6, and then using a more familiar form of the Boltzmann principle such as that given in equation 2-27.

A modulus of the form $E(t) = E_0 e^{-t/\tau}$ corresponds to a compliance $D(t) = 1/E_0 + t/E_0\tau$. Thus, using equation (2-27) we get:

$$
\epsilon(t) = \int_{-\infty}^{t} \frac{d\sigma(u)}{du} D(t-u)\,du
$$

$$
\epsilon(t) = \sigma_0 D(t) + \int_{0}^{t'} \frac{d\sigma(u)}{du} D(t-u)\,du - \sigma(t')D(t-t')
$$

The first and last terms arise from the fact that step stresses of σ_0 at 0 and $-\sigma(t')$ at t' are applied to the sample. When $0 < u < t'$, $\sigma(u) = \epsilon_0 E_0 e^{-u/\tau}$ so that $d\sigma(u)/du = -(\epsilon_0 E_0/\tau)e^{-u/\tau}$ and

$$
\epsilon(t) = \sigma_0\left(\frac{1}{E_0} + \frac{t}{E_0\tau}\right) + \int_{0}^{t'}\left(-\frac{\epsilon_0 E_0}{\tau}\right)e^{-u/\tau}\left[\frac{1}{E_0} + \frac{t-u}{\tau E_0}\right]du
$$

$$
- \epsilon_0 E_0 e^{-t'/\tau}\left[\frac{1}{E_0} + \frac{t-t'}{\tau E_0}\right]
$$

Recognizing that $\sigma_0/E_0 = \epsilon_0$, factorization of ϵ_0 gives

$$\epsilon(t) = \epsilon_0\left[\left(1 + \frac{t}{\tau} - \left(1 + \frac{t-t'}{\tau}\right)e^{-t'/\tau} - \int_0^{t'} e^{-u/\tau}\left(1 + \frac{t}{\tau} - \frac{u}{\tau}\right)\frac{du}{\tau}\right.\right]$$

$$= \epsilon_0\left[\left(1 + \frac{t}{\tau} - \left(1 + \frac{t-t'}{\tau}\right)e^{-t'/\tau} - \int_0^{t'/\tau} e^{-x}\,dx - \frac{t}{\tau}\int_0^{t'/\tau} e^{-x}\,dx\right.\right.$$

$$+ \int_0^{t'/\tau} x e^{-x}\,dx\Bigg]$$

$$= \epsilon_0\left[\left(1 + \frac{t}{\tau} - \left(1 - \frac{t-t'}{\tau}\right)e^{-t'/\tau} + e^{-x}\Big|_0^{t'/\tau} + \frac{t}{\tau}e^{-x}\Big|_0^{t'/\tau} - e^{-x}(1+x)\Big|_0^{t'/\tau}\right]$$

$$\epsilon(t) = \epsilon_0(1 - e^{-t'/\tau})$$

CHAPTER 3

3-1(a) Equation (3-6) states

$$\log a_T = \frac{-C_1(T - T_g)}{C_2 + T - T_g} \tag{3-6}$$

The apparent energy of activation for viscoelastic relaxation may be defined as

$$\Delta H_a = R\frac{d\ln a_T}{d(1/T)}$$

which by simple calculus is just

$$\Delta H_a = -RT^2(2.303)\frac{d\log a_T}{dT} \tag{a}$$

Using equation (a) on equation (3-6) yields

$$\Delta H_a = \frac{2.303\,C_1 C_2 RT^2}{(C_2 + T - T_g)^2} \tag{b}$$

$$\frac{(C_2 + T - T_g)(-C_1) - \{-C_1(T - T_g)\}}{(C_2 + T - T_g)^2}$$

$$= \frac{(-C_1 C_2) - \cancel{}{} + C_1 T_g - \{-C_1 T}{(C_2 + T - T_g)^2} + C_1 T_g\}$$

$$= \frac{C_1 C_2 RT^2(2.303)}{(C_2 + T - T_g)^2}$$

Since we are interested in the activation energy at T_g, equation (b) is just

cal
mole

$$\Delta H_a = 2.303 \left(\frac{C_1}{C_2} \right) R T_g^2$$

Using the universal constants given in Table 3-2, at 200°K this quantity is about 61 kcal mole^{-1}.

(b)   As $T \gg T_g$ equation (b) becomes

$$\Delta H_a = 2.303 R C_1 C_2$$

which is equal to 4.1 kcal mole^{-1}.

3-2.   From Equation (3-19), we have

$$C_1 = \frac{B}{2.303} f_g \quad \text{and} \quad C_2 = \frac{f_g}{\alpha_f}$$

where the reference temperature has been taken as T_g. At any other reference temperature, say $T°$, we have

$$C_1° = \frac{B}{2.303(f_g + \alpha_f(T_0 - T_g))}$$

and

$$C_2° = \frac{f_g + \alpha_f(T_0 - T_g)}{\alpha_f}$$

where equation (3-16) has been used. Algebraic manipulation of these four equations gives

$$C_1° = \frac{C_1}{1 + (T_0 - T_g)/C_2}$$

and

$$C_2° = C_2 + T_0 - T_g$$

Utilizing the universal constants, at $T_0 = T_g + 50°C$,

$$C_1° = 8.83 \quad \text{and} \quad C_2° = 101.6° C$$

3-3.   This problem is worked out in reference 9 of Chapter 3.

3-4.   Starting with the arbitrary temperature $T = T_g - 5°C$, substitution into equation (3-6) with values of C_1 and C_2 from Table 3-2 gives

$$\log a_T = -1.87 \quad \text{or} \quad a_T = 0.0135$$

Thus, at this temperature

$$\tau_i(T_g - 5) = \frac{\tau_i(T_g)}{a_T}$$

and the two relaxation times become

$$\tau_1(T_g - 5) = \frac{1\,\text{sec}}{0.0135} = 7.4 \times 10^1\,\text{sec}$$

$$\tau_2(T_g - 5) = \frac{10^4\,\text{sec}}{0.0135} = 7.4 \times 10^5\,\text{sec}$$

Thus,

$$E(10\,\text{sec}, T_g - 5) = (3.0 \times 10^9 e^{-10\,\text{sec}/74\,\text{sec}} + 5.0 \times 10^5 e^{-10\,\text{sec}/(7.4 \times 10^5 \text{sec})})\,Pa$$

$$= 2.6 \times 10^9\,Pa$$

Modulus values at other temperatures are calculated in the same way.

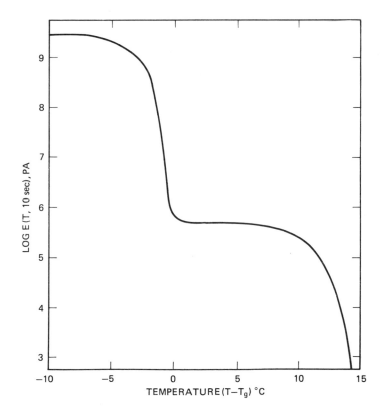

3-5.

$$G(t, T) = \sum_{i=1}^{n} G_i e^{-t/\tau_i(T)}$$

$$G(t, T_r) = \sum_{i=1}^{n} G_i e^{-t/\tau_i(T_r)}$$

The correspondence principle says

$$G\left(\frac{t}{a_T}, T\right) = G(t, T_r)$$

or

$$\sum_{i=1}^{n} G_i e^{-t/a_T\tau_i(T)} = \sum_{i=1}^{n} G_i e^{-t/\tau_i(T_r)}$$

For this to be true for all values of t, each individual exponent must be equal for all values of t:

$$e^{-t/a_T\tau_i(T)} = e^{-t/\tau_i(T_r)} \qquad 1 \leqslant i \leqslant n$$

or

$$\frac{t}{a_T\tau_i(T)} = \frac{t}{\tau_i(T_r)}$$

$$a_T = \frac{\tau_i(T_r)}{\tau_i(T)} = \frac{A_i \, e^{H/RT_r}}{A_i \, e^{H/RT}}$$

$$\ln a_T = \frac{H}{R}\left(\frac{1}{T_r} - \frac{1}{T}\right)$$

$$\log a_T = \frac{H}{2.3R}\left(\frac{1}{T_r} - \frac{1}{T}\right)$$

3-6. From equation (3-6):

$$\log a_T = \frac{C_1(T - T_g)}{C_2 + T - T_g}$$

which may be rearranged to give

$$\log a_T = \frac{-C_1(T - T_g + C_2)}{T - T_g + C_2} + \frac{C_1 C_2}{T - T_g + C_2}$$

$$= -C_1 + \frac{C_1 C_2}{T - T_g + C_2}$$

If:

$$\beta = C_1$$

$$\alpha = \frac{1}{C_1 C_2}$$

$$T_0 = T_g - C_2$$

we have the Vogel expression.

CHAPTER 4

4-1. Equation (3-16) may be written as

$$f_M = f_g + \alpha_f(T - T_{gM})$$

$$f_\infty = f_g + \alpha_f(T - T_{g\infty})$$

where f_M is the fractional free volume at some temperature and molecular weight. Here we have assumed α_f and f_g are both independent of molecular weight. These relationships may be introduced into the equation given in the problem to yield equation (4-28).

4.4 (a) The weight fraction of chain ends, w_e, is just the weight of the chain ends divided by the total weight of the polymer. If n_i is the number of moles of polymer with molecular weight M_i, w_e is given as

$$w_e = \frac{\text{weight of chain ends}}{\text{total weight}} = \frac{\sum_i n_i M_e}{\sum_i n_i M_i} = M_e\left(\frac{\sum_i n_i}{\sum_i n_i M_i}\right)$$

The term in parentheses, however, is just the reciprocal of the number average molecular weight of the polymer, $\overline{M}_n$, so

$$\overline{M}_n = \frac{M_e}{w_e}$$

Substitution of this relationship into equation (4-2) yields the desired result.

(b) Rewriting the equation given in terms of relevant parameters, remembering that $w_p + w_e = 1$, gives

$$T_g = T_g^\infty (1 - w_e) + T_{ge} w_e + K w_e - K(w_e)^2$$

Since w_e is small for polymers, the last term can be neglected and

$$T_g = T_g^\infty - w_e (T_g^\infty - T_{ge} - K)$$

This expression is identical to the desired result if we identify the term in parentheses with c/M_e.

4-5. The reference here should be sufficient.

4-6.

$$\frac{d\delta}{dt} = -\frac{\delta}{\tau} \qquad \text{where } \tau \text{ does not depend on } \delta$$

Thus:

$$\frac{d\delta}{\delta} = d \ln \delta = -\frac{dt}{\tau}$$

$$\int_{\delta = \delta_0}^{\delta = \delta(t)} d \ln \delta = \int_0^t \frac{dt}{\tau}$$

$$\ln \delta(t) - \ln \delta_0 = -\frac{t}{\tau}$$

$$\delta(t) = \delta_0 e^{-t/\tau}$$

With the parameters given:

$$\delta(t) = 3 \times 10^{-3} e^{-t/\text{hour}}.$$

This function is plotted in the accompanying figure.

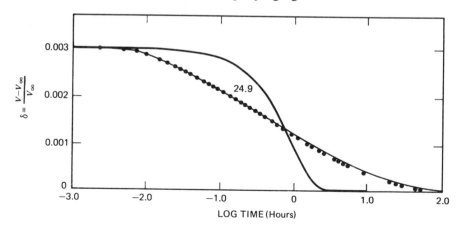

CHAPTER 5

5-1.

$$\overline{r^2} = \frac{1}{1} \sum_{i=1}^{4} (r_i)^2 = (r_1)^2$$

$$\overline{r^2} = \left(\sum_{j=1}^{4} I_j \right)^2 \tag{5-2}$$

$$\begin{aligned}
\overline{r^2} = \quad & I_1 \cdot I_1 + I_1 \cdot I_2 + I_1 \cdot I_3 + I_1 \cdot I_4 + \\
& I_2 \cdot I_1 + I_2 \cdot I_2 + I_2 \cdot I_3 + I_2 \cdot I_4 + \\
& I_3 \cdot I_1 + I_3 \cdot I_2 + I_3 \cdot I_3 + I_3 \cdot I_4 + \\
& I_4 \cdot I_1 + I_4 \cdot I_2 + I_4 \cdot I_3 + I_4 \cdot I_4
\end{aligned}$$

5.3. Equation (5-32) may be written as

$$\overline{r^2} = \frac{\int_0^\infty r^4 e^{-b^2 r^2} \, dr}{\int_0^\infty r^2 e^{-b^2 r^2} \, dr}$$

after using equation (5-30) to eliminate $\omega(r) \, dr$. This expression integrates to

$$\overline{r^2} = \frac{3}{2b^2}$$

Since b^2 is $3/2\overline{r^2}$ [equation (5-29)], we get $\overline{r^2} = \overline{r^2}$.

5-4. This is just an extension of problem 5-1. Using the answers given there,

$$\begin{aligned}
\overline{r^2} = \quad & l^2 \left[1 + \cos \theta + \cos^2 \theta + \cos^3 \theta \right] + \\
& l^2 \left[\cos \theta + 1 + \cos \theta + \cos^2 \theta \right] + \\
& l^2 \left[\cos^2 \theta + \cos \theta + 1 + \cos \theta \right] + \\
& l^2 \left[\cos^3 \theta + \cos^2 \theta + \cos \theta + 1 \right]
\end{aligned}$$

5-5. Equation (5-43) states

$$n_e = \frac{R^2}{\overline{r^2}} \quad \text{and} \quad l_e = \frac{\overline{r^2}}{R}$$

If the length of each monomer is 4.6Å, R must be

$$R = 4.6\text{Å}n$$

where n is the number of monomer units. Thus,

$$n_e = \frac{(4.6n)^2}{16.2n} = 1.3n$$

and

$$l_e = \frac{16.2n}{4.6n} = 3.52\text{Å}$$

5-6.   Equation (5-16) reads

$$\omega(m, n) = \left(\frac{1}{2}\right)^n \frac{n!}{[(n+m)/2]![(n-m)/2]!} \tag{5-16}$$

and if $n \gg m$, we can then use the form of Stirling's approximation given in the problem to get

$$\ln \omega(m, n) = \left(n + \frac{1}{2}\right)\ln n - \frac{1}{2}(n + m + 1)\ln\left[\left(\frac{n}{2}\right)\left(1 + \frac{m}{n}\right)\right]$$

$$- \frac{1}{2}(n - m + 1)\ln\left[\left(\frac{n}{2}\right)\left(1 - \frac{m}{n}\right)\right]$$

$$- \frac{1}{2}\ln 2\pi - n \ln 2$$

However, since $n \gg m$, we can use the series expansion

$$\ln\left(1 \pm \frac{m}{n}\right) = \pm \frac{m}{n} - \frac{m^2}{2n^2}$$

where third-order and higher-order terms of (m/n) are neglected. We now have:

$$\ln \omega(m, n) = \left(n + \frac{1}{2}\right)\ln n - \frac{1}{2}\ln 2\pi - n \ln 2 - \frac{1}{2}(n + m + 1)$$

$$\times \left(\ln n - \ln 2 + \frac{m}{n} - \frac{m^2}{2n^2}\right) - \frac{1}{2}(n - m + 1)$$

$$\times \left(\ln n - \ln 2 - \frac{m}{n} - \frac{m^2}{2n^2}\right)$$

Simplifying and taking the antilogarithm of both sides, we obtain

$$\omega(m, n) = \left(\frac{2}{\pi n}\right)^{1/2} \exp\left(-\frac{m^2}{2n}\right)$$

In the simplification we have made use of the fact that $n \gg 1$, so that

$$\frac{m^2}{2n}\left(\frac{1}{n} - 1\right) \approx -\frac{m^2}{2n}$$

5-7. Equation (5-32) is restated in this case as

$$\overline{r^2}' = \frac{\displaystyle\int_{1/2b}^{\infty} r^4 e^{-b^2 r^2}\, dr}{\displaystyle\int_0^{\infty} r^2 e^{-b^2 r^2}\, dr}$$

where $\overline{r^2}'$ indicates a particular average quantity. Our problem then becomes one of evaluating the definite integrals given. This equation may be rewritten as

$$\overline{r^2}' = \frac{\displaystyle\int_0^{\infty} r^4 e^{-b^2 r^2}\, dr - \int_0^{1/2b} r^4 e^{-b^2 r^2}\, dr}{\displaystyle\int_0^{\infty} r^2 e^{-b^2 r^2}\, dr - \int_0^{1/2b} r^2 e^{-b^2 r^2}\, dr}$$

Now consider the second term in both the numerator and the denominator of this expression. If r is between 0 and $1/2b$, the exponential is quite close to 1. Thus we may approximate this expression as

$$\overline{r^2}' = \frac{\displaystyle\int_0^{\infty} r^4 e^{-b^2 r^2}\, dr - \int_0^{1/2b} r^4\, dr}{\displaystyle\int_0^{\infty} r^2 e^{-b^2 r^2}\, dr - \int_0^{1/2b} r^2\, dr}$$

which is integrated to give

$$\overline{r^2}' = \frac{3}{2b^2}\left[\frac{\sqrt{\pi} - \frac{1}{60}}{\sqrt{\pi} - \frac{1}{6}}\right]$$

Since the expression in brackets is greater than 1.0, $\overline{r^2}' > \overline{r^2}$. This result is clearly expected since conformations with small $\overline{r^2}$ have been excluded in calculating $\overline{r^2}'$.

5-8. Once again equation (5-32) is the starting point.

$$\bar{r} = \frac{\displaystyle\int_0^\infty r^3 e^{-b^2 r^2}\, dr}{\displaystyle\int_0^\infty r^2 e^{-b^2 r^2}\, dr}$$

In this case the variable transformation

$$y = b^2 r^2$$

is helpful. This yields

$$\bar{r} = \frac{1}{b}\frac{\displaystyle\int_0^\infty y e^{-y}\, dy}{\displaystyle\int_0^\infty y^{1/2} e^{-y}\, dy}$$

Both of these integrals are complete gamma functions so that

$$\bar{r} = \frac{1}{b}\frac{\Gamma(2)}{\Gamma(\frac{3}{2})} = \frac{1}{b}\frac{1\cdot\Gamma(1)}{\frac{1}{2}\Gamma(\frac{1}{2})}$$

or

$$\bar{r} = \frac{2}{b\sqrt{\pi}}$$

In comparing the text with the results of this calculation we see that

$$\left(\overline{r^2}\right)^{1/2} \neq \bar{r}$$

In addition, one should recall that $\bar{r}$ for a Gaussian chain is zero. These results are not inconsistent, since what makes the vector average zero is its sign: just as many **r** are positive as are negative along any axis. In this problem, we have calculated the average of the radius of a spherical shell, a scalar quantity that cannot take on negative values.

5-9(a) From equation (5-56):

$$f = 2kTb^2\mathbf{r}$$

Thus:

$$P = \frac{f}{A} = \frac{2kTb^2\mathbf{r}}{A}$$

In our case,

$$P = \frac{2kTb^2 \cdot 1/b}{S^2} = \frac{2kTb}{S^2} = \frac{2kT}{S^2L} = \frac{2kT}{V}$$

(b)

$$100 \text{ atm} = \frac{(2)(.0821 \; l \text{ atm}/\text{mole}°K)(300°K)(1/N_A)}{(10 \times 10^{-16}\text{cm}^2)(1 \times 10^{-6}\text{cm})}$$

$$T = 366°K$$

CHAPTER 6

6-1(a)

$$dH = T\,dS + V\,dP + f\,dL$$

Thus:

$$\left(\frac{\partial H}{\partial L}\right)_{T,P} = T\left(\frac{\partial S}{\partial L}\right)_{T,P} + f$$

The Maxwell relation

$$\left(\frac{\partial S}{\partial L}\right)_{T,P} = -\left(\frac{\partial f}{\partial T}\right)_{L,P}$$

gives

$$\left(\frac{\partial H}{\partial L}\right)_{T,P} = -T\left(\frac{\partial f}{\partial T}\right)_{L,P} + f$$

Now

$$\left(\frac{\partial f}{\partial T}\right)_{P,L} = \left(\frac{\partial f}{\partial T}\right)_{P,\lambda} + \left(\frac{\partial f}{\partial \lambda}\right)_{P,L}\left(\frac{\partial \lambda}{\partial T}\right)_{P,L}$$

Evaluating the various coefficients using equation (6-60) gives

$$\left(\frac{\partial f}{\partial T}\right)_{P,L} = \frac{f}{T} - \frac{\alpha f}{3}\frac{\lambda^3 + 2}{\lambda^3 - 1}$$

and

$$\left(\frac{\partial H}{\partial L}\right)_{T,P} = \frac{\alpha Tf}{3}\frac{\lambda^3 + 2}{\lambda^3 - 1}$$

(b)   Analogous to part (a) starting with dU instead of dH.

(c)

$$\left(\frac{\partial H}{\partial L}\right)_{T,P} = \frac{T\alpha}{\beta}\left(\frac{\partial V}{\partial L}\right)_{T,P}$$

Since

$$\left(\frac{\partial U}{\partial L}\right)_{T,V} = 0$$

Substituting the result of part (a):

$$\left(\frac{\partial V}{\partial L}\right)_{P,T} = \frac{\beta f}{3}\frac{\lambda^3 + 2}{\lambda^3 - 1}$$

Note that since $(\partial H/\partial L)_{T,P} \neq 0$, an ideal rubber has a thermoelastic inversion point.

6-2.   Equation (6-11) defines the stress–temperature coefficient in terms of pertinent quantities.

$$\left(\frac{\partial f}{\partial T}\right)_{P,L} = \frac{f - (\delta H/\delta L)_{T,P}}{T}$$

In problem 6-1a,

$$\left(\frac{\partial H}{\partial L}\right)_{T,P} = \frac{\alpha Tf}{3}\frac{\lambda^3 + 2}{\lambda^3 - 1}$$

So

$$\left(\frac{\partial f}{\partial T}\right)_{P,L} = \frac{f}{T}\left[1 - \frac{\alpha T}{3}\frac{\lambda^3 + 2}{\lambda^3 - 1}\right]$$

6-3. Equation (6-42) in terms of this problem states

$$\Delta A = \frac{1}{3} NkT \overline{b^2 r^2} \left(\lambda^2 + \frac{1}{\lambda^2} - 2 \right)$$

Incorporation of (6-44) gives

$$\Delta A = \frac{1}{2} NkT \frac{\overline{r_0^2}}{\overline{r_f^2}} \left(\lambda^2 + \frac{1}{\lambda^2} - 2 \right)$$

Since pure shear is a constant volume process ($\lambda_1 \lambda_2 \lambda_3 = 1$),

$$\alpha^* = \lambda$$

and equation (6-52b) becomes

$$f = \frac{1}{L} \left(\frac{\partial A}{\partial \lambda} \right)_{T,V}$$

Operating with this on the second expression gives:

$$f = \frac{NkT}{L} \frac{\overline{r_0^2}}{\overline{r_f^2}} \left(\lambda - \frac{1}{\lambda^3} \right)$$

or

$$\sigma = N_0 kT \frac{\overline{r_0^2}}{\overline{r_f^2}} \left(\lambda - \frac{1}{\lambda^3} \right)$$

6-4. According to equation (6-74)

$$I_1 = \lambda^2 + 1 + \frac{1}{\lambda^2}$$

and

$$I_2 = \lambda^2 + 1 + \frac{1}{\lambda^2}$$

Thus equation (6-80) states:

$$\overline{W} = C_{100} \left(\lambda^2 + \frac{1}{\lambda^2} - 2 \right) + C_{010} \left(\lambda^2 + \frac{1}{\lambda^2} - 2 \right)$$

and using equation (6-79) gives

$$\sigma = 2(C_{100} + C_{010})\left(\lambda - \frac{1}{\lambda^3}\right)$$

6-6. Substitution into equation (6-84) gives

$$G_0 = \frac{(0.95\text{g/cm}^3)(8.3 \times 10^7\text{erg/mole°K})(300°\text{K})}{5000\text{g/mole}}\left(1 - \frac{10^4}{10^5}\right)$$

$$G_0 = 4.3 \times 10^6\text{dyne/cm}^2 = 4.3 \times 10^5\text{Pa}.$$

Here we have assumed that the front factor $\overline{r_0^2}/\overline{r_f^2}$ is equal to unity.

6-7. We are told that the sample ruptures at 100% extension in the unswollen state, that is, at $\lambda = 2$. To achieve this strain by swelling an isotropic rubber, we must have $\lambda_x = \lambda_y = \lambda_z = 2$. At this point, $V_r = \frac{1}{8}$ since $V_r = \lambda_x\lambda_y\lambda_z^{-1}$. Thus the sample's volume is expected to increase by a factor of 8 before rupture.

6-8. Using equation (6-94)

$$E_f = \left[1 + 2.5(0.3) + 14.1(0.3)^2\right]5 \times 10^7\text{dynes/cm}^2$$

$$E_f = 1.5 \times 10^8\text{dynes/cm}^2$$

If half of the filler particles become ineffective,

$$V_f = 0.15$$

and

$$E_f = 8.3 \times 10^7\text{dynes/cm}^2$$

6-9. Starting with equation (6-66)

$$f = G_0 A_0\left(\lambda - \frac{V}{V_0\lambda^2}\right)$$

one can write

$$f = G_0 A_0\left(\frac{L}{L_0} - \frac{V}{V_0}\frac{L_0^2}{L^2}\right)$$

Differentiation yields

$$\left(\frac{\partial f}{\partial T}\right)_{V,L} = \left(\frac{\partial G_0}{\partial T}\right)_{V,L} A_0 \left(\frac{L}{L_0} - \frac{V}{V_0}\frac{L_0^2}{L^2}\right)$$

$$+ \left(\frac{\partial A_0}{\partial T}\right)_{V,L} G_0 \left(\frac{L}{L_0} - \frac{V}{V_0}\frac{L_0^2}{L^2}\right)$$

$$+ G_0 A_0 \left[-\frac{L}{L_0^2}\left(\frac{\partial L_0}{\partial T}\right)_{V,L} + \frac{V}{V_0^2}\frac{L_0^2}{L^2}\left(\frac{\partial V_0}{\partial T}\right)_{V,L} \right.$$

$$\left. -2\frac{V}{V_0}\frac{L_0}{L^2}\left(\frac{\partial L_0}{\partial T}\right)_{V,L} \right]$$

Grouping terms and using the approximation

$$\frac{1}{V_0}\left(\frac{\partial V_0}{\partial T}\right)_{V,L} = \frac{3}{2}\frac{1}{A_0}\left(\frac{\partial A_0}{\partial T}\right)_{V,L} = 3\frac{1}{L_0}\left(\frac{\partial L_0}{\partial T}\right)_{V,L}$$

leads directly to equation (6-67).

CHAPTER 7

7-1. Figure 7-3b is a plot of the relaxation behavior of a Maxwell body on a log–log scale. Thus we want to calculate

$$\frac{d\ln E(t)}{d\ln t}$$

where

$$E(t) = Ee^{-t/\tau}$$

However,

$$\ln E(t) = \ln E - \frac{t}{\tau}$$

and since

$$\frac{d \ln E(t)}{d \ln t} = \frac{t d \ln E(t)}{dt}$$

$$\frac{d \ln E(t)}{d \ln t} = -\frac{t}{\tau}$$

which approaches $-\infty$ for $t \gg \tau$.

7-2 and 7-3. As an example, we will work with $E'(\omega)$ for a Maxwell body:

(a) Equation (2-56) gives

$$E'(\omega) = \omega \int_0^\infty \sin \omega s E(s) \, ds$$

$$E'(\omega) = \omega \int_0^\infty \sin \omega s E e^{-s/\tau} \, ds$$

which integrates to

$$E'(\omega) = \frac{E \omega^2 \tau^2}{1 + \omega^2 \tau^2}$$

(b) Starting with equation (7-5) we find:

$$E \epsilon_0 \int_{t_1}^{t_2} e^{t/\tau} \cos \omega t \, dt = \int_{\sigma(t_1)e^{t_1/\tau}}^{\sigma(t_2)e^{t_2/\tau}} d(\sigma e^{t/\tau})$$

assuming that

$$\epsilon(t) = \epsilon_0 \sin \omega t$$

This expression integrates to

$$\sigma(t_2) - \sigma(t_1)e^{t_1 - t_2/\tau} = \frac{E \omega \epsilon_0}{1/\tau^2 + \omega^2} \left[\frac{1}{\tau} \cos \omega t_2 + \omega \sin \omega t_2 \right.$$

$$\left. - e^{t_1 - t_2/\tau} \left(\frac{1}{\tau} \cos \omega t_1 + \omega \sin \omega t_1 \right) \right]$$

Since our result must be independent of the starting time t_1, and the present time t_2 ($E'(\omega)$ is not a function of t) let $t_2 \to \infty$. Then

$$\sigma(t_2) = \frac{E\omega\epsilon_0}{1/\tau^2 + \omega^2}\left(\frac{1}{\tau}\cos \omega t_2 + \omega \sin \omega t_2\right) \tag{1}$$

$E'(\omega)$ however measures only the in-phase stress, so

$$E'(\omega) = \frac{\sigma'(t_2)}{\epsilon(t_2)} = \frac{E\omega\epsilon_0}{1/\tau^2 + \omega^2}\left(\frac{\omega \sin \omega t_2}{\epsilon_0 \sin \omega t_2}\right)$$

or

$$E'(\omega) = \frac{E\omega^2\tau^2}{1 + \omega^2\tau^2}$$

7-4. Here we can start with equation (1) of example 7-2 with added subscripts.

$$\sigma_i(t_2) = \frac{E_i\omega\epsilon_0}{1/\tau_i^2 + \omega^2}\left(\frac{1}{\tau_i}\cos \omega t_2 + \omega \sin \omega t_2\right)$$

This expression can be separated into an in-phase σ_i' part and an out-of-phase σ_i'' part, as follows:

$$\sigma_i'(t_2) = \frac{E_i\omega\epsilon_0}{1/\tau_i^2 + \omega^2}(\omega \sin \omega t_2)$$

$$\sigma_i''(t_2) = \frac{E_i\omega\epsilon_0}{1/\tau_i^2 + \omega^2}\left(\frac{1}{\tau_i}\cos \omega t_2\right)$$

Now the total in-phase stress considering all elements is just

$$\sigma'(t_2) = \sum_i \frac{E_i\omega^2\epsilon_0}{1/\tau_i^2 + \omega^2}(\sin \omega t_2)$$

Here the quantity in parentheses merely expresses the phase of the response (in phase) and we may write immediately:

$$E'(\omega) = \sum_i \frac{E_i\omega^2}{1/\tau_i^2 + \omega^2} = \sum_i \frac{E_i\omega^2\tau_i^2}{1 + \omega^2\tau_i^2}$$

Similarly, for the out-of-phase response,

$$\sigma''(t_2) = \sum_i \frac{E_i \omega \epsilon_0}{1/\tau_i^2 + \omega^2} \left(\frac{1}{\tau_i} \cos \omega t_2 \right)$$

Here $\cos \omega t_2$ just stipulates out-of-phase response and

$$E''(\omega) = \sum_i \frac{E_i \omega \tau_i}{1 + \omega^2 \tau_i^2}$$

Then according to equation (2-22) written for tension:

$$E^*(\omega) = \sum_i \frac{E_i \omega^2 \tau_i^2}{1 + \omega^2 \tau_i^2} + i \sum_i \frac{E_i \omega \tau_i}{1 + \omega^2 \tau_i^2}$$

Clearly such sums are difficult to invert analytically.

7-5.   Here we can write

$$\frac{d\epsilon_1}{dt} = \frac{1}{E_1} \frac{d\sigma_1}{dt} + \frac{\sigma_1}{\eta_1}$$

$$\frac{d\epsilon_i}{dt} = \frac{1}{E_i} \frac{d\sigma_i}{dt} + \frac{\sigma_i}{\eta_i}$$

$$\frac{d\epsilon_z}{dt} = \frac{1}{E_z} \frac{d\sigma_z}{dt} + \frac{\sigma_z}{\eta_z}$$

These expressions simplify considerably, since:

$$\sigma_1 = \sigma_2 \cdots = \sigma_i = \cdots \sigma_z$$

and

$$\sum_{i=1}^{z} \frac{d\epsilon_i}{dt} = \frac{d\epsilon}{dt}$$

Thus,

$$\frac{d\epsilon}{dt} = \frac{d\sigma}{dt} \sum_{i=1}^{z} \frac{1}{E_i} + \sigma \sum_{i=1}^{z} \frac{1}{\eta_i}$$

Defining

$$\frac{1}{\mathcal{E}} = \sum_{i=1}^{z} \frac{1}{E_i}$$

and

$$\frac{1}{H} = \sum_{i=1}^{z} \frac{1}{\eta_i}$$

this equation becomes:

$$\frac{d\epsilon}{dt} = \frac{1}{\mathcal{E}} \frac{d\sigma}{dt} + \frac{\sigma}{H}$$

which is solvable in the usual way.

7-6(a) Equation (7-70) can be rearranged to give

$$\rho a^2 z^2 = \frac{36}{N}(\eta)$$

Since all of the terms on the left side are parameters relating to the model itself and not to the specific experimental conditions, this result may be substituted into the expression for τ_p in the problem to give the desired result.

(b) Table 7-2 gives an expression for the tensile creep compliance of a Voigt–Kelvin Model as

$$D(t) = \sum_{p=1}^{z} D_p(1 - e^{-t/\tau_p})$$

Comparison of this expression with that in the problem yields the result that

$$D_p = \frac{8}{3NkT\pi^2} \frac{1}{p^2}$$

7-7. First let us investigate the values of $E''(\omega)$ and ω at the maximum of a $\log E''(\omega)$ versus $\log \omega$ curve for a single Maxwell element.

$$\log E''(\omega) = \log E + \log \omega + \log \tau - \log(1 + \omega^2\tau^2)$$

$$\frac{d\log E''(\omega)}{d\log \omega} = 1 - \frac{2\omega^2\tau^2}{1 + \omega^2\tau^2}$$

At the maximum this slope is zero and ω is found to equal $1/\tau$. Substitution of this result into the defining equation for $E''(\omega)$ shows that

$$E''(\omega) = \frac{E}{2}$$

at the maximum. Thus,

$$\tau_1 = 10^{-2} \quad \text{and} \quad E_1 = 4 \times 10^6$$

$$\tau_2 = 10^{-9} \quad \text{and} \quad E_2 = 2 \times 10^{10}$$

7-8.

$$\eta = \int_0^\infty G(t)\, dt \tag{3-7}$$

For a Maxwell–Wiechert model:

$$\eta = \int_0^\infty \sum_{i=1}^z G_i e^{-t/\tau_i}\, dt$$

$$\eta = \sum_{i=1}^z G_i \int_0^\infty e^{-t/\tau_i}\, dt$$

$$= \sum_{i=1}^z G_i \tau_i$$

7-9.

$$E(t) = \sum_{p=1}^z E_p e^{-t/\tau_p}$$

$$\approx E_p \int_{p=1}^z e^{-tp/\tau_{max}}\, dp$$

$$\approx E_p \frac{\tau_{max}}{t} \left(e^{-t/\tau_{max}} - e^{-t/\tau_{min}} \right)$$

If $\tau_{min} < t < \tau_{max}$, the quantity in parentheses is close to 1 and relatively constant. Thus,

$$\frac{d \ln E(t)}{d \ln t} = -1$$

7-10.   Equation (7-35) defines $H(\tau)$ as

$$E(t) = \int_{\ln \tau = -\infty}^{\ln \tau = \infty} H(\tau) e^{-t/\tau} d \ln \tau \qquad (7\text{-}35)$$

For $\tau \ll t$, the exponential is 0.0 whereas for $\tau \gg t$, it is 1.0. If the exponential is approximated by a step function from 0.0 to 1.0 at $t = \tau$, we have

$$E(t) \approx \int_{\ln \tau = \ln t}^{\ln \tau = \infty} H(\tau) d \ln \tau$$

Furthermore,

$$E(t + \Delta) - E(t) \approx -\int_{\ln \tau = \ln t}^{\ln \tau = \ln t + \Delta} H(\tau) d \ln \tau \approx H(\tau) \Big|_{t = \tau} (\ln t + \Delta - \ln t)$$

or, in the limit $\Delta \to 0$,

$$\frac{- dE(t)}{d \ln t} \Big|_{t = \tau} \approx H(\tau) \equiv H_1(\tau)$$

7-11(a)

$$E(t) = E_0 e^{-t/\tau_m}$$

where the subscript m is included to remind us that the relaxation time is constant:

$$\frac{dE(t)}{d \ln t} = \frac{t \, dE(t)}{dt} = -\frac{t E_0 e^{-t/\tau_m}}{\tau_m}$$

$$H_1(\tau) \equiv \frac{dE(t)}{d \ln t} \Big|_{t = \tau} = \frac{\tau}{\tau_m} E_0 e^{-\tau/\tau_m}$$

which is plotted below.

(b) $\dfrac{d^2E(t)}{d(\ln t)^2} = \dfrac{t\,d(dE(t)/d\ln t)}{dt} = t\left(-\dfrac{E_0}{\tau_m}e^{-t/\tau_m} + \dfrac{tE_0}{\tau_m^2}e^{-t/\tau_m}\right)$

$H_2(\tau) = \left(E_0\dfrac{t}{\tau_m}e^{-t/\tau_m} - E_o\dfrac{t}{\tau_m}e^{-t/\tau_m} + \dfrac{t^2}{\tau_m^2}E_0e^{-t/\tau_m}\right)_{t=2\tau}$

$= 4E_0\left(\dfrac{\tau}{\tau_m}\right)^2 e^{-2\tau/\tau_m}$

which is also plotted in the figure.

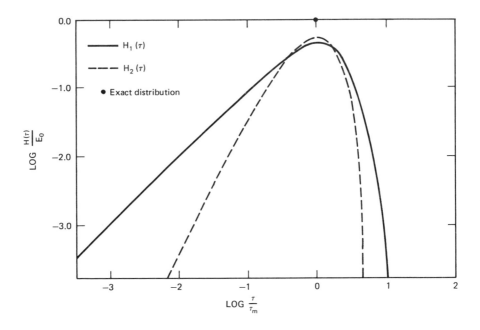

CHAPTER 8

8-1. For an electrical circuit in parallel, the charges in the various branches add, the voltages are everywhere the same:

$$Q = Q_1 + Q_2$$

From equation (8-9),

$$Q_1 = C_1 V_0 (1 - e^{-t/\tau})$$

and

$$Q_2 = C_2 V_0$$

Therefore:

$$Q(t) = C_1 V_0 (1 - e^{-t/\tau}) + C_2 V_0$$

or:

$$C(t) = C_1 (1 - e^{-t/\tau}) + C_2$$

Using the definitions of C_R and C_u and substituting

$$C(t) = (C_R - C_u)(1 - C^{-t/\tau}) + C_u$$

8-2(a)   This is entirely analogous to the derivation of the mean square end to end distance of the freely jointed chain given in Chapter 5, Section A. The result is:

$$\langle \mu^2 \rangle = N \mu_0^2$$

(b)   Following the argument for the dimensions of a chain with tetrahedral valence angles given in Chapter 5, Section C, the result is:

$$\langle \mu^2 \rangle = 2 N \mu_0^2$$

Since the bonds of this chain are free to rotate about one another (bond rotational potentials are zero), $g = 1$.

(c)   If the chain possesses barriers to rotation about backbone bonds, g will take on a value different from 1 since the cosine averages in equation 8-55 are now different from 0.

(d)   We must calculate the cosine averages for the angles of each bond with respect to the first. Since we are given that the rotation potentials of each bond are independent and each bond can only assume the angles $\pi/2$,

π, and $3\pi/2$ with respect to the first, we have:

$$\langle \cos \gamma_{1j} \rangle = \frac{\cos \pi/2 + \cos \pi + \cos 3\pi/2}{3} = -\frac{1}{3}$$

Thus $g = 0.67$

$$\langle \mu^2 \rangle = 2gN\mu_0^2 = 1.34N\mu_0^2$$

8-3. Let $a = \epsilon_R - \epsilon_u$; $b = \epsilon_R + \epsilon_u$; $X = \omega\tau$

$$\left(\epsilon_u + \frac{a}{1+X^2} - \frac{b}{2}\right)^2 + \left(\frac{aX}{1+X^2}\right)^2 = \frac{a^2}{4}\left(\frac{1-X^2}{1+X^2}\right)^2 + a^2\left(\frac{X}{1+X^2}\right)^2$$

$$= a^2\frac{\left(1-X^2\right)^2 + 4X^2}{4\left(1+X^2\right)^2} = \left(\frac{\epsilon_R - \epsilon_u}{2}\right)^2$$

8-4.

$$q^* = C^*V^*, \qquad C^* = C_0\epsilon^*$$

$$V^*(t) = V_0\exp(i\omega t), \qquad \epsilon^* = \epsilon' - i\epsilon''$$

$$I^* = \frac{dq^*}{dt} = C^*\frac{dV^*}{dt} = (i\omega C^*)V^* = Y^*V^*$$

Y^* (is admittance) $= i\omega C^* = \omega C_0(\epsilon'' + i\epsilon')$. For a series of R's and C's,

$$Y^* = \frac{1}{R_s - i/\omega C_s}$$

By comparing real and imaginary terms one has:

$$\epsilon'' = \frac{R_s\omega C_s^2}{C_0\left(R_s^2\omega^2 C_s^2 + 1\right)} \qquad \epsilon' = \frac{C_s}{C_0\left(R_s^2\omega^2 C_s^2 + 1\right)}$$

8-5. Consider only that part of I^* which is in phase with V^*. From Problem 8-4 one has: $I^* = i\omega C^*V^* = C_0\omega\epsilon''V^* + i(C_0\omega\epsilon')V^*$

$$\text{Power} = V^* \cdot I^* = C_0\omega\epsilon''|V^*|^2 = C_0V_0^2\omega\epsilon''$$

Since

$$\epsilon'' = \frac{(\epsilon_R - \epsilon_u)\omega\tau}{1 + \omega^2\tau^2},$$

$$P = \left[C_0 V_0^2 (\epsilon_R - \epsilon_u) \right] \frac{\omega^2\tau}{1 + \omega^2\tau^2}$$

8-6.   As can be shown by methods of Chapter 7,

$$D'' = \frac{\omega\tau}{E_2(1 + \omega^2\tau^2)} \left.\begin{array}{l} \\ \\ \\ \\ \\ \\ \end{array}\right\} \qquad \text{where } \tau = \frac{\eta}{E_2}$$

$$E_1 = \text{separate spring}$$

$$D' = \frac{1}{E_1} + \frac{1}{E_2(1 + \omega^2\tau^2)} \qquad E_2 = \text{Voigt spring}$$

To complete the analogy,

$$\frac{1}{E_2} = \epsilon_R - \epsilon_u; \qquad \frac{1}{E_1} = \epsilon_u$$

CHAPTER 9

9-2.   Binomial expansion yields

$$\left(1 - \frac{q}{N(0)x}\right)^x = 1 - \frac{xq}{N(0)x} + \frac{(x)(x-1)}{2!}\left(\frac{q}{N(0)x}\right)^2$$

$$- \frac{(x)(x-1)(x-2)}{3!}\left(\frac{q}{N(0)x}\right)^3 \cdots$$

$$\approx 1 - \frac{q}{N(0)} + \frac{1}{2!}\left(\frac{q}{N(0)}\right)^2 - \frac{1}{3!}\left(\frac{q}{N(0)}\right)^3$$

which is just the expansion of the exponential $e^{-q/N(0)}$. Thus

$$\left(1 - \frac{q}{N(0)x}\right)^x \approx e^{-q/N(0)}$$

9-3.   Start with equation (9-10), which states

$$q = - N(0)\ln\frac{\sigma(t)}{\sigma(0)}$$

The ln function can be expanded according to the series

$$\ln x = \frac{x-1}{x} + \frac{1}{2}\left(\frac{x-1}{x}\right)^2 + \frac{1}{3}\left(\frac{x-1}{x}\right)^3 \cdots$$

But if x is close to 1

$$\ln x \approx \frac{x-1}{x}$$

Thus:

$$q \approx - N(0)\left(\frac{\sigma(t)/\sigma(0)-1}{\sigma(t)/\sigma(0)}\right)$$

$$q = - N(0)\left(\frac{\sigma(0)}{\sigma(t)}-1\right) \qquad\qquad (9\text{-}20)$$

9-4.   This can be done by simply writing several terms of the binomial expansion

$$\left(1-\frac{1}{N(0)}\right)^q = 1 - \frac{q}{N(0)}$$

$$+ \frac{q(q-1)}{2!}\left(\frac{1}{N(0)}\right)^2 - \frac{q(q-1)(q-2)}{3!}\left(\frac{1}{N(0)}\right)^3 \cdots$$

9-5.   This summation may be written explicitly as

$$N(0)p\left[1+2(1-p)+3(1-p)^2+4(1-p)^3+\cdots\right]$$

Consider the series in brackets and replace $1-p$ by y. Thus we desire the sum

$$1+2y+3y^2+4y^3+\cdots$$

The MacLaurin expansion of $1/(1-y)^2$ generates just this sum. Working back through the variable transform,

$$N(0)p\frac{1}{p^2} = \frac{N(0)}{p}$$

the result quoted in equation (9-28).

9-6. From equation (9-37),

$$\frac{N_u}{N_x}\left[\left(\frac{L_s}{L_u}\right)^2 - \frac{L_u}{L_s}\right] = \frac{L_x}{L_s} - \left(\frac{L_s}{L_x}\right)^2$$

the following rearrangements can be made:

$$\frac{N_u}{N_x}\left[\frac{L_s^3 - L_u^3}{L_u^2 L_s}\right] = \frac{L_x^3 - L_s^3}{L_x^2 L_s}$$

$$\frac{N_u L_x^2}{N_x L_u^2} = \frac{(L_x^3/L_u^3) - (L_s^3/L_u^3)}{(L_s^3/L_u^3) - 1} = \frac{(L_x^3/L_u^3) - 1}{(L_s^3/L_u^3) - 1} - 1$$

$$\frac{L_s^3}{L_u^3} - 1 = \frac{(L_x^3/L_u^3) - 1}{(N_u L_x^2/N_x L_u^2) + 1}$$

$$\frac{L_s}{L_u} = \left[\frac{(L_x^3/L_u^3) - 1}{(N_u L_x^2/N_x L_u^2) + 1} + 1\right]^{1/3}$$

The permanent set is defined by equation (9-33):

$$\text{p.s.} = \frac{L_s - L_u}{L_x - L_u} \times 100$$

$$= \frac{(L_s/L_u) - 1}{(L_x/L_u) - 1} \times 100$$

$$= \frac{100}{(L_x/L_u) - 1}\left\{\left[\frac{(L_x^3/L_u^3) - 1}{(N_u L_x^2/N_x L_u^2) + 1} + 1\right]^{1/3} - 1\right\}$$

$$= C_3\left\{\left[\frac{C_1}{(N_u/N_x)C_2 + 1} + 1\right]^{1/3} - 1\right\}$$

where C_1, C_2, and C_3 are defined as in the text.

9-7. In general a rate constant can be written as

$$k = Ae^{-H_a/RT}$$

where all of the temperature dependence is embodied in the exponential term. If one performs the operation

$$RT^2 \frac{d \ln k}{dT}$$

the activation energy is extracted. Applying this procedure to equation (9-50) yields (9-51).

List of Major Symbols

Certain symbols in this book were used to denote different meanings. This redundancy is preferred in order to preserve conformity with those commonly accepted in the literature. Those symbols whose usage is restricted to certain chapters are so indicated. Some minor symbols are defined in the text where applicable and are not included in this list.

A	Helmholtz free energy
A	normalization constant (Chapter 8)
A_0	cross-sectional area
$[A]$	Zimm matrix
a	entanglement parameter
a	atomic radii (Chapter 8)
a^2	mean square end-to-end distance of submolecule
a_s	radius of the dielectric sphere
a_T	shift factor
B	$3kT/a^2\rho$ (Chapter 7)
b^2	$3nl^2/2$
C	capacitance (Chapter 8)
C_0	capacitance in vacuum
C_p	constant pressure heat capacity
C_V	constant volume heat capacity
C_r	reference capacitance
C_s	sample capacitance
C_1, C_2	WLF parameters
C_1, C_2	Mooney–Rivlin parameters (Chapter 6)

C_1, C_2, C_3	permanent set parameters (Chapter 9)
c	concentration
c	Fox–Flory constant (Chapter 4)
D	diffusion constant (Chapter 8)
D	tensile compliance
$D(t)$	tensile creep compliance
d	density
d	capacitor plate spacing (Chapter 8)
E	electric field (Chapter 8)
E, E_0	tensile modulus
$E(t)$	tensile relaxation modulus
E^*	complex tensile modulus
$E^*(\omega, t)$	time-dependent periodic electric field (Chapter 8)
E'	dynamic storage tensile modulus
E'	internal electric field (Chapter 8)
E''	dynamic loss tensile modulus
E_e	equilibrium tensile modulus
E_f	tensile modulus of a filled rubber
E_1	glassy tensile modulus
E_2	rubbery tensile modulus
e	charge of the electron
e_s	efficiency of scission
F	Gibbs free energy
F	General function (Chapter 2)
f	force
f	fractional free volume (Chapter 3)
f_e	energetic stress
f_g	fractional free volume at glass transition
f_s	entropic stress
G, G_0	shear modulus
$G(t)$	shear relaxation modulus
G^*	complex shear modulus
G'	dynamic storage shear modulus
G''	dynamic loss shear modulus
G_1	glassy shear modulus
G_2	rubbery shear modulus
g	Kirkwood correlation factor
H	enthalpy
H_a	activation energy
$H(\tau)$	distribution of relaxation times
I	current
I_1, I_2, I_3	strain invariants

$[I]$	identity matrix
I_0	initiator concentration
i	$\sqrt{-1}$
J	shear compliance
$J(t)$	shear creep compliance
J^*	complex shear compliance
J'	dynamic storage shear compliance
J''	dynamic loss shear compliance
J_r	recoverable compliance
K_A, K_B	Gordon–Taylor parameters
k	Boltzmann constant
k	K_A/K_B (Chapter 4)
k_d	rate constant for initiator decomposition
k_i	rate constant for initiation
k_p	rate constant for propagation
k_t	rate constant for termination
L	tube length (Chapter 7)
L	length of strained sample
L	Laplace transform (Chapter 2)
L_0	length of unstrained sample at V_0
L'	length of unstrained sample at V
L_s	new unstrained length
L_u	original unstrained length
L_x	strained length
$L(\tau)$	distribution of retardation times
l	length of a chain link
$\vec{l}$	vectorial length of a chain link
M	modulus (Chapter 1)
M	concentration of molecular segments (Chapter 9)
M	molecular weight
M_c	molecular weight of network chain (Chapter 6)
M_c	critical entanglement molecular weight (Chapter 7)
M_n	number average molecular weight
M_s	total electric moment in a dielectric sphere
m	difference between positive and negative steps (Chapter 5)
m	number of polysulfide links (Chapter 9)
N	number of network chains
N_A	Avogadro's number (Chapter 8)
N_{Av}	Avogadro's number
N_0	number of network chains per unit volume
N_s	number of molecules in a dielectric sphere
N_u	original chain concentration

N_x	chain concentration corresponding to permanent set length
n	fixed number of conformations of chain
n	number of links in a polymer chain
n_+	number of positive steps
n_-	number of negative steps
P	pressure
P_D	long time limit of the time dependent portion of the polarization
P_R	infinite time polarization
P	instantaneous polarization
$P(t)$	time-dependent polarization
$P_D(t)$	time-dependent portion of the polarization
$P^*(\omega)$	complex dielectric constant
p	size parameter
Q	charge (Chapter 8)
$[Q]$	transformation matrix
q	number of cuts per unit volume (Chapter 9)
q	elements of $[q]$ matrix (Chapter 7)
$\dot{q}$	time derivative of q
$[q]$	normal coordinate matrix
R	resistance (Chapter 8)
R	extended chain length (Chapter 5)
R	ideal gas constant
R_r	reference resistance
R_s	sample resistance
r	charge separation (Chapter 8)
$\overline{r^2}$	mean square end-to-end distance
$\vec{r}$	vectorial length of chain ends
$\overline{r_f^2}$	mean square end-to-end distance of a free chain
$\overline{r_0^2}$	mean square end-to-end distance of a real chain
S	entropy
s	negative slope in the transition region of modulus–temperature curve
T	temperature
T_g	glass transition temperature
T_i	inflection temperature
T_m	melting temperature
T_{ref}	reference temperature
T_2	second order transition temperature
t	time
t_{ref}	reference time

U	internal energy
u_0	hole-formation energy
V	volume
V	voltage (Chapter 8)
V_c	capacitive potential
V_d	volume fraction of diluent
V_f	free volume
V_f	volume fraction of filler (Chapter 6)
V_0	initial volume
V_p	volume fraction of polymer
V_R	resistive potential
V_r	volume fraction of rubber
V_s	voltage across a sample
V_∞	equilibrium volume
v	velocity
W	work
$\overline{W}$	strain energy
W_A, W_B	weight fraction of comonomers
X	calculational variable (Chapter 8)
X	displacement
$\dot{X}$	time derivative of displacement
X_A, X_B	mole fractions of comonomers
x	number of segments per chain
x	diffusion distance (Chapter 7)
Y	calculational variable (Chapter 8)
Z	number of primary valences of chain atom
z	number of mechanical elements or submolecules
α	thermal expansion coefficient
α	effective polarizability (Chapter 8)
α^*	extension ratio at constant volume
α_f	thermal expansion coefficient of free volume
α_g	thermal expansion coefficient of glass
α_0	molecular polarizability
α_s	polarizability of a dielectric sphere
α_r	thermal expansion coefficient of rubber
β	isothermal compressibility
β_f	isothermal compressibility of free volume
Γ	gamma function
γ	shear strain
$\vec{\gamma}$	strain vector
γ'	in-phase component of strain vector
γ''	out-of-phase component of strain vector

γ_{ij}	angle between ith and jth chain segment (Chapter 8)
Δ	logarithmic decrement
δ	loss angle
δ	normalized volume departure from equilibrium (Chapter 4)
ε	tensile strain
ε	dielectric constant (Chapter 8)
$\varepsilon, \varepsilon_A, \varepsilon_B$	flex energies (Chapter 4)
ε_R	long time limit of the dielectric constant
ε_u	instantaneous dielectric constant
ε'	real part of the complex dielectric constant
ε''	imaginary part of the complex dielectric constant
$\varepsilon^*(\omega)$	complex dielectric constant
η	viscosity
θ	$\pi - \psi$
Λ	extension ratio including strain amplification factor
$[\Lambda]$	diagonalized Zimm matrix
λ	extension ratio at constant pressure
λ_c	extension ratio of dry rubber
λ_p	elements of $[\lambda]$ matrix
λ_s	extension ratio of swollen rubber
$\lambda_1, \lambda_2, \lambda_3$	principal extension ratios
μ	Poisson ratio
μ	molecular mobility (Chapter 7)
μ	dipole moment (Chapter 8)
π	180° angle or 3.1416
ρ	density (Chapter 3)
ρ	segmental friction factor (Chapter 7)
ρ_0	segmental friction factor at short times
$\vec{\sigma}$	stress vector
$\vec{\sigma}'$	in-phase component of stress vector
$\vec{\sigma}''$	out-of-phase component of stress vector
σ_s	shear stress
σ_t	tensile stress
σ'	true stress
σ	charge density (Chapter 8)
σ_0	charge density in vacuum
$\sigma_1, \sigma_2, \sigma_3$	principal stresses
τ	relaxation or retardation time
τ^*	effective relaxation time
τ_c	critical relaxation time
$\tau_{\min}$	minimum relaxation time

τ_p	relaxation time of Rouse model
τ_1	maximum relaxation time
$[\phi]$	transformation matrix
ψ	bond angle
Ω	number of conformations
ω	angular frequency (Chapters 2, 7)
ω	probability (Chapters 5, 6)
ζ	ordering parameter

Author Index

Subject Index

*Index prepared by Helmut G. Hermann